Galileo's Pendulum

SUNY series in Science, Technology, and Society
Sal Restivo and Jennifer Croissant, editors

**Property of CEL
Lancaster University
2004**

GALILEO'S PENDULUM

*Science, Sexuality,
and the Body-Instrument Link*

Dušan I. Bjelić

Foreword by Michael Lynch

State University of New York Press

Published by
State University of New York Press, Albany

© 2003 State University of New York

All rights reserved

Printed in the United States of America

No part of this book may be used or reproduced in any manner whatsoever without written permission. No part of this book may be stored in a retrieval system or transmitted in any form or by any means including electronic, electrostatic, magnetic tape, mechanical, photocopying, recording, or otherwise without the prior permission in writing of the publisher.

For information, address State University of New York Press,
90 State Street, Suite 700, Albany, NY 12207

Production by Judith Block
Marketing by Anne Valentine

Library of Congress Cataloging-in-Publication Data

Bjelic, Dusan I.
   Galileo's pendulum : science, sexuality, and the body-instrument link / Dusan I. Bjelic.
     p. cm. — (SUNY series in science, technology, and society)
   Includes bibliographical references and index.
   ISBN 0-7914-5881-4 (hc : alk. paper) — ISBN 0-7914-5882-2 (pbk. : alk. paper)
    1. Science—Methodology. 2. Ethnomethodology. 3. Pendulum. 4. Galilei, Galileo, 1564–1642. I. Title. II. Series.

Q175.B566 2003
501—dc22

                                         2003058124

10 9 8 7 6 5 4 3 2 1

For Nahod

Contents

Foreword by Michael Lynch	ix
Acknowledgments	xv
Introduction	1

PART ONE. PLEASURE

1.	Time, Pleasure, and Knowledge	19
2.	The Perversion of Objectivity and the Objectivity of Perversion	33
3.	The Jesuits' Homosocial Ties and the Experiments with Galileo's Pendulum	59

PART TWO. PEDAGOGY

4.	The "Body-Instrument Link" and the Prism: A Case Study	81
5.	The Formal Structure of Galileo's Pendulum	115
6.	The Respecification of Galileo's Pendulum	129
	Conclusion	153
	Notes	161
	Index	199

Foreword

Michael Lynch

To appreciate Dušan Bjelić's treatment of Galileo's pendulum it is necessary to understand what it is not. It is not a straightforward contribution to the history of science. Nor is it a "social" or "philosophical" study of science in the usual sense. The book is sociologically inspired, it investigates the fundamental philosophical question of the relationship between experience and the world, and it includes some instructive forays into the history of Western science, but the book's historical, philosophical, and sociological treatments do not easily fit within the usual taxonomies of academic disciplines, subdisciplines, and interdisciplines. Instead, this book is a highly imaginative—and yet exquisitely material—investigation of the embodied practice of demonstrating "Galilean" science.

I can imagine that many readers will be appalled by very idea that Galileo's pendulum is an expression of "sexuality." Those of us familiar with the recent "science wars"—which fortunately seem to be on the wane—will recall that self-appointed defenders of science and reason seized, almost gleefully, upon statements in the science studies literature suggesting that Western science is an exercise in rape or that the experimental method is "gendered" at its core and in every detail. Those who are inclined to seek evidence of degenerative thinking about science in the social sciences and humanities will find plenty of grist for their mill in this book. The overall thesis is perverse, the writing is peculiar, and the history is quirky. Superficially, the book is a blasphemous exercise in the history and philosophy of science. However, readers who dismiss the book out of hand, who settle for harsh denunciations of the author's twisted mind, or who read it in search of outlandish sentences in order to illustrate charges of irrationality and incompetence, will have missed the serious and original lessons offered by this book.

Bjelić writes with tongue in cheek: Galileo's "sexuality" is not what you might think; indeed, "Galileo" is "himself" an elusive figure in this text whose mythical demonstration of universal laws is mimed in Bjelić's performance. Bjelić also has something serious to say, and the challenge offered by his book is to find the serious lesson while appreciating the laughing tone of voice with which it is enunciated. Bjelić relishes strangeness, and yet speaks perversely about the most ordinary of matters: the mundane practice of *doing* a classical scientific demonstration. In the context of classic history and philosophy of science—with its hagiographic tendencies and grand metaphysical narratives—nothing is stranger than the ordinary way in which science is performed.

Galileo's Pendulum not only describes an historic experiment (or, in the Koyrean view, a theorized projection of an experiment), it is itself a textual experiment. The chapters on Galileo's pendulum demonstration invite readers to perform what the text says: to gather string and weights, assemble an apparatus, and play with the assemblage in search of the Galilean law. Later chapters invite readers to pick up a toy prism in order to generate the geometrical phenomena the text describes. The text should be read in concert with the relevant embodied performance. If Bjelić could have managed to do so economically, he would have supplied a kit of materials with this book, so that readers would have ready-to-hand the required equipment for "reading" his descriptions. The idea of building an exercise around a Galilean demonstration derives from Harold Garfinkel, the "founding father" of ethnomethodology. Garfinkel (2002, ch. 9) recently published an account of an exercise that he and a few colleagues performed in the 1980s. The exercise involved an attempt to reenact Galileo's inclined plane demonstration of speed in the law of falling bodies. (Whether or not it was "Galileo's" or "Garfinkel's" inclined plane demonstration is an open question that, for reasons I shall elaborate, is beside the point.) Garfinkel describes an exercise performed in an office at UCLA with a twenty-two foot board purchased from a lumberyard. The motive for the exercise was not to gain insight into a foundational historical moment in the annals of physical science. Garfinkel and his colleagues did not try to reproduce exactly what Galileo actually may have done four centuries ago (assuming that he ever did perform the inclined plane demonstration). Instead, the exercise was a way of gaining first-hand insight into a foundational phenomenon in the annals of *social* science. Garfinkel built and performed an improvised version of Galileo's demonstration as a way of "making and describing workplace ordinary *social facts* of practical action and practical reason" (Garfinkel 2002, 264).

To some extent, Garfinkel's Galileo derives from Husserl's (1970) imaginative treatment of "Galileo" as a foundational personage who encapsulates all of the tendencies that Husserl ascribes to modem mathematical science. "Galileo" puts forward a universal mathematical law with minimal explication of the embodied practices through which the law is made concretely accountable. But whereas Husserl asks how the law arises from the phenomenological experience that its reified statement obscures, Garfinkel asks how the law is locally performed, intersubjectively, in practice. He does not doubt the truth or demonstrability of the law of falling bodies, but instead seeks to recover and respecify the local, embodied practices that make the law come true, as an *accountable universal law of motion*. Garfinkel's Galileo is twinned with Garfinkel's Durkheim: the physical law is a social fact, and neither the law nor the fact is adequately expressed in positive idioms. In its classic formulation, the Galilean law is invariant to the conditions of its performance; it is part of the *Mathesis Universalis*. In its classic formulation, the Durkheimian social fact is an "immortal" phenomenon: the social orders described by such facts persist despite changes in the local staff who reproduce those orders; they are external to individual wills and coercive in the way they force individuals to comply with them (Durkheim 1982 [1895]). Much in the way that Husserl demands a phenomenological explication of the situated experiential conditions that give rise to and sustain the universality of the physical law, Garfinkel demands an ethnomethodological demonstration of just how an "immortal" social fact is achieved, locally and relentlessly, in a specific setting of conduct. The specific setting of conduct happens to be an attempt to reproduce "Galileo's" inclined plane demonstration, and the social facts in question are part and parcel of the achieved regularity and demonstrable rigor of the demonstration. By critically explicating Durkheim through a performance of Galileo's demonstration, Garfinkel and his local work crew attempt to respecify two birds with one stone. Both Galileo's physical law and Durkheim's social fact are (re)produced together: they are intertwined with, and yet practically teased out from, a local, concerted achievement. The universality of the law and the immortality of the fact reside in a reflexively grasped ability to go on in the "same" way, indefinitely, through changes in staff, variations in equipment, and encounters with "demonic" contingencies.[1]

Bjelić was inspired by Garfinkel's treatment of Galileo's demonstration. He first learned of it from Garfinkel during a meeting in Boston in the late 1980s. At the time,[2] Bjelić did not get from Garfinkel an elaborate account of the demonstration or of the ethnomethodological rationale for it. He had to work out for himself what might be so interesting and promising about ethnomethodological respecifications of classic demonstrations. He took up another of Galileo's demonstrations—the demonstration with three pendulums of the inverse square relationship between the length of drop and period of swing.[3] This demonstration was more manageable than the inclined plane experiment (no twenty-two foot board is required). Bjelić also worked with optical demonstrations using prisms. In one instance, he developed a way to demonstrate Goethe's (claimed) refutation of Newton's optics. A reader/viewer of the demonstration was required to retrace an argument by viewing a series of figures reproduced in an article (Bjelić & Lynch 1992). The figures were designed after the "cards" in Goethe's "color games," and the play between prism and cards made up a distinctive form of visual argument. Bjelić also developed the novel set of exercises for using a prism presented in chapter 4. In that demonstration a geometrical order is produced through a distinctive practice that exploits refractive and reflective surfaces of a prism.

As historical research on Galileo demonstrates, there is no single way to work out what Galileo may have actually done with the demonstrations he described. Galileo's descriptions of the inclined plane and pendulum demonstrations are thin on detail and leave much to the imagination. Some historians—most notably Stillman Drake (1975)—set themselves the task of recovering what Galileo may, or even must, have done with the materials available to him. Without denying the ingenuity of Drake's and others' attempts at historical recovery, Bjelić attempts a very different project. He does not attempt to replicate Galileo's demonstrations.[4] Like Garfinkel's exercise, Bjelić's low-budget effort to re-do Galileo's demonstration shows endless differences between the "original" and the "replication." Such demonstrations are not, cannot be, and do not intend to be "faithful" renderings of Galileo's original actions.

There is a sense, however, in which Bjelić's descriptions and instructions invite readers faithfully to reproduce Galileo's *phenomenon*. The reproduction is not a matter of duplicating original materials and executing the same actions—something that would be impossible in any case—but of acting in accord with a set of instructions and successfully "getting" the phenomenon the instructions promise to deliver. Galileo provides sketchy instructions about the particular pendulum demonstration Bjelić attempts to perform. The instructions do not specify what kind of string or bob to use; Galileo's ratios between the lengths of the strings (which are, in fact, inaccurate) do not specify the scale and composition of the device from which to mount the apparatus. A great deal is left to work out by the future "cohorts" of Galileo's readers. As Bjelić describes, one can experiment with different types of string and bob. If one needs to come up with materials quickly, rather than ordering a ready made device from a scientific equipment supply house, it can be difficult to devise a frame, or to find objects of the "right" weight and size that are easy to attach to the strings. Not everything will serve the purpose of demonstrating the universal law, but readers may discover that many things will serve in a pinch. The vagueness of the instructions and the necessity to invent or discover how to follow them are *part* of the phenomenon. Despite, or even because of, the vague and open-ended instructions, it is not difficult to perform a successful demonstration of the law Galileo describes. Indeed, a peculiarity of this demonstration is that it is so *robust*. One can devise an open series of substitutions of materials and methods and still get the demonstration to work.[5] It is far more difficult to succeed, for example, with descriptions of nineteenth century electrical experiments, or even with standard high school chemistry experiments.[6]

So what is Bjelić's phenomenon? What does he hope to demonstrate? The mathematical "law" with which Galileo describes, and accounts for, the demonstrable relation between the periodicity and length of a pendulum is *not* Bjelić's phenomenon. Bjelić does not provide an anti- realist debunking of the law; instead, he situates the law within the practice of demonstrating it. The law is an integral and intelligible part of the demonstration. The phenomenon for Bjelić is not the disembodied law, but the way the law is contingently bound to the embodied performance and witnessing of the details of the demonstration. Bjelić's investigation is not merely an exercise in following instructions, because the question of whether one has followed the instructions or not turns on the contingent visibility of the law.

Bjelić's book is an ethnomethodological study, which in the earlier chapters is presented by means of a Lacanian voice. I am not a Lacanian, or for that matter an enthusiast for any brand of psychoanalytic theory, and so I tend to read Bjelić's text in terms of the less overtly sexualized phenomenological treatments of the body and embodied action. Although some readers no doubt will be put off by Bjelić's sexualized treatment of the Galilean exercise, others may derive more satisfaction—not to speak of pleasure—from them. However, even if one concludes that talk of Galileo's "sexuality" is a gratuitous play on words, it should be possible to read this book not only as a source of information, but also as an installation that facilitates an unusually direct, material engagement with foundational issues in the history, philosophy, and sociology of physical science.

REFERENCES

Bjelić, Dušan, and Michael Lynch (1992). "The work of a scientific demonstration: Respecifying Newton's and Goethe's theories of color," pp. 52–78 in G. Watson and R. Seiler (eds.), *Text in Context: Contributions to Ethnomethodology*. London and Beverly Hills: Sage.

Collins, H. M. (1985). *Changing Order: Replication and Induction in Scientific Practice*. London and Beverly Hills: Sage.

Drake, Stillman (1975). "The role of music in Galileo's experiments," *Scientific American*, 232 (6): 98–104.

Durkheim, Emile (1982 [1895]). *The Rules of Sociological Method*. S. Lukes (trans.). New York: Free Press.

Garfinkel, Harold (2002). *Ethnomethodology's Program: Working out Durkheim's Aphorism*. London: Rowman & Littlefield.

Garfinkel, Harold, Eric Livingston, Michael Lynch, Douglas Macbeth, and Albert B. Robillard (1989). "Respecifying the natural sciences as discovering sciences of practical action (I and II): Doing so ethnographically by administering a schedule of contingencies in discussions with laboratory scientists and by hanging around their laboratories," Unpublished Paper (Department of Sociology, UCLA).

Husserl, Edmund (1970). *The Crisis of European Sciences and Transcendental Phenomenology: An Introduction to Phenomenological Philosophy*. David Carr (trans.). Evanston, Ill.: Northwestern University Press.

Livingston, Eric (1995). "The idiosyncratic specificity of the methods of physical experimentation," *Australia and New Zealand Journal of Sociology* (ANZJS), 31 (3): 1–22.

Acknowledgments

For writing the first part of the book I would like to extend my gratitude to Lucinda Cole, who gave me a fresh perspective on the writings of Michel Foucault and gave me a generous help in the composition of the text. Discussions with Arnold Davidson about Foucault, Mario Biagioli, and James MacLachlan on Galileo were extremely helpful in the narrative invention of "Foucault's pendulum." I also thank Marcus Hellyer for his insights into the history of the Jesuits. For the second part of the book, I would like to thank to my "ethnosquad," David Bogen, Eric Livingston, Michael Lynch, and Doug MacBeth, who spent a great deal of time experimenting and discussing various aspects of the relation between the body and scientific instrumentation. They were all as passionately involved in the project as I was. Shelley Schweizer, James Shay, Kenneth Rosen, and Rosemary Miller must be thanked for their generous help in improving various parts of my text. Carl Hass should be thanked for the illustrations. Acknowledgment also goes to Jean-Claude Sabrier for allowing me to reproduce his photograph of the erotic Swiss watch and to The University of Chicago Press for allowing me to reprint their published materials. The figures in chapters 4 and 5 previously appeared in *Human Studies*, volumes 18 (1995) and 19 (1996), published by Kluwer Academic Publishers B.V., and were redrawn for this book. I am also thankful to SUNY Press's anonymous readers for their constructive criticism.

Introduction

"People used to have sex at these conferences," a prominent cultural theorist of science from an Ivy League university nostalgically observed at an international conference on science. "Now," she laments, "all they do is to talk about their work." Her lamentation seems to miss here a larger point about the growing commerce between rationality and sex, between sexual deprivation on the one hand and scientific productivity on the other. In a not altogether unrelated story, two other "professionals of a rational discourse," as chance would have it, had less reason to lament. In an e-mail message that came to my attention, he, a historian of science, writes to her, a literary scholar, about *his* female graduate student being sexually propositioned by a male professor at her first academic conference. "She will finally understand," he sarcastically concludes "what a mind is all about." Indeed our historian of science and our literary scholar enjoyed the fruits of their correspondence at their next conference, where even long-term relationships with other individuals could not deprive them of erotic pleasures with each other. In retrospect his e-mail is nothing so much as a veiled sexual proposition toward her. What the complaining scholar did not understand seems to come naturally to our historian of science. Rather than being explicit about his own desire toward our literary scholar, he writes about others' sexual machinations and pleasures. In other words, he deploys an ironic textual courtship, a technique of writing as having "desire" of others through language.[1] His message, potent with erotic improvisation and implicit sexual suggestion, creates a desire that may or may not ever be carnally realized. If it is realized then it is caused by the text and the preceding techniques of writing. If it is not realized then it remains as a creative desire for the production of discursive rationality, professional bonding, career moves, or rekindling a stagnated research. The above lamented conference, deprived of carnal pleasures, might have been a situation of an intense professional sexualization, a productive erotic competitiveness in the service of advancing discursive rationality.

In other words, the lesson learned here is that sexual deprivation in the production of scientific rationality not only does not exclude pleasure, but the sexualization of science is a professional requirement for its competitive productivity. Consider, for instance, how science deprived of sexual pleasures resexualizes the object of its inquiry in order to produce itself in the arena of a discourse of rationality in the case of the science of primates. Discussing those who scientifically study primates, Donna Haraway observes that "these people have engaged in dynamic, disciplined, and intimate relations of love and knowledge with animals they were privileged to watch."[2] Birute Galdikas, sometimes referred to as "the Jane Goodall of orangutans" and the world's foremost authority on these great apes, when suddenly seized by a large male orangutan with an erect phallus, tells her companion not to worry because the beast has "a very small penis."[3] "In the eyes of someone who has lived much of her life with orangutans," Peter Singer comments on this event, "to be seen by one of them as an object of sexual interest is not a cause for shock or horror,"[4] because our rational inquiry into the life of primates, he observes, both intimizes our relations with them and changes the old normative boundaries between humans and animals.[5] What remains the same is that we maintain "disciplined and intimate relations to love and knowledge"[6] as general conditions of research and, as Foucault would have claimed, this proves that in modern science "sex" is a general method of truth.

Are, then, scientists lustless eunuchs, or seekers of perverse pleasures? Is science a discipline of the mind, or a new *aphrodisiac?* Both sets of questions make sense, if we take into account Foucault's claim that throughout the history of Western society, knowledge has burgeoned out of one's relationship to pleasure. This specific Western brand of knowledge had, unlike in the Orient, Foucault postulates, very little to contribute to human sexuality except what he calls a discursive sexuality, or a "pleasure of analysis."[7] Considering this point, one may argue that in the invention and the production of science, the Western discourse on sexuality found its self-inventing continuum.

I shall argue this point by using Foucault's history of sexuality as a framework for re-examining the history of seventeenth-century experimental physics. On the trail of Nietzsche's "will to power," Foucault held the view that strategies of domination, "the 'show' of strength or superior force . . ."[8] create new cultural values, engender history itself, and become the power of historical innovation. Since Foucault moved away in his late work from a judicial model of power, for him as for Nietzsche, creative power meant primarily strategies of self-domination and self-hegemony, or what the Greeks called *askēsis*. As techniques of depriving oneself of pleasures, *askēsis* for the Greeks was not a way to suppress pleasures but rather a rational choice to resexualize these very techniques of deprivation and intensify pleasures by keeping them suspended as a "perpetual erotic possibility" for new relations of domination.⁹ Rather than eroticizing just the genitals, the Greeks eroticized the entire body and its public appearance through measured conduct, gaining in turn not only domination over their own bodies and pleasures but also over those of their fellow citizens. This creative force then is the self coming into the world. The self which must "always overcome itself"¹⁰ and in the process innovate and create new values, is, Foucault insists, rationality born of awareness of pleasures concealed as a hidden and a creative force underneath a judicial model of power. Here then, Judith Butler concludes, "Productive power in the sexual domain thus becomes understood as a kind of erotic improvisation, the sexual version of Nietzsche's life-affirming creation of values."¹¹

Adhering to this Foucaultian model of the relation between pleasure and rationality, self and knowledge, I shall, throughout this book, construct the history of seventeenth-century experimental physics in order to argue that since Plato's definition of knowledge as pure desire and the introduction of chastity into the method of seventeenth-century natural philosophy from Neoplatonic and Christian asceticism, Western science falls under the rubric of Butler's "erotic improvisation." An erotic domination creative of seventeenth-century scientific innovations had specific modalities of its relation to pleasure, power, and the self that need to be revealed. While Greek strategies of self-domination were centered in the self, seventeenth-century natural philosophy centered its strategies of self-domination within monastic or scientific brotherhoods. The basic form of power revolved, then, around subjugation to the various rules and prohibitions that these brotherhoods imposed upon the body of a natural philosopher with regard to pleasures primarily towards women, as a prerequisite for trustworthy scientific research. Here conditions of collective sexual deprivation operated both as self-fashioning but also as an epistemic condition of scientific method, accentuating Foucault's claim that with Enlightenment "sex" became a general method of securing knowledge. Such a constellation of the techniques of sexual deprivation and of scientific research among men, gave rise to homosociality among natural philosophers, a primal bonding among men as a mode of self-empowerment and domination over women as well as the male's body and pleasures.¹² Its homoerotic desire became a creative force for scientific research, for historical innovation and the new history of scientific rationality.¹³

Infused with domination through knowledge, homosocial creativity and the innovation of seventeenth-century science through "erotic improvisation" also had an oppressive dimension. Its sexual hierarchies that arise from homosocial bonding and homoerotic desires, resemble, to a large extent, ancient Greek marriage. In Greek marriage the woman is allowed to have sexual relations only with her husband, and only because she is under his control; on the other hand, for the husband, having sexual relations with his wife is "the most elegant way of exercising [that] control."¹⁴ While chastity was a political necessity for women, it was an "aesthetic flourish for men."[15] This sadomasochistic model of Greek power relations, in which women are subjugated to the eroticism of men—the source of flourishing for the men themselves—seems to operate as a productive framework for seventeenth-century science. While women abstained from science because it was a political necessity, men flourished and aestheticized themselves as a new brand of scientist who now might, with instruments such as Galileo's pendulum, unify the Earth and the skies as they themselves are unified, homoerotically. With their homoerotic rationality, seventeenth-century scientists transformed the old matter into the purified matter of a machine, and created a new world on the principles of intensified sexual exclusion of women from rationality and homoerotic pleasure.

The commerce of this economy of pleasure seems to escape the mainstream philosophy and history of science, which was preoccupied either with science as pure rationality or as an effect arising from the judicial mode. If we agree with Foucault's fresh view on rationality, that the invention of discursive pleasures became *a force* manufactured by the new scientific subjectivity, we should then disagree with Edmund Husserl's attitude toward "Galilean science" as a self-contained model of positive reasoning, characteristic of an all-modern natural science that is independent of pleasure, power, and subjectivity. Closer to our view, then, is the work of Mario Biagioli and his celebrated claim that Galileo's science came out of professional self-fashioning, a combination of controlled epistemic and courtly procedures.[16] To support Biagioli's claim, we may paraphrase Ian Hunter by saying that "Galilean science" is a set of autonomous ethical techniques and practices by which scientists must unavoidably problematize their pleasures (historically and discursively defined as "sexuality").[17] As a prerequisite for the proliferation of scientific rationality, science as a method of self-fashioning forces scientists like Galileo to "conduct themselves as the subjects of an aesthetic existence" in order to be innovative.[18]

Taking all of this into account makes our job of understanding the complexities of the relation between scientific subjectivity, domination and knowledge in seventeenth-century science all the more complex. As a homosocial form of domination, the science ushered in by Galileo simultaneously oppresses and affirms an identity. While Galileo fashioned himself as a new brand of natural philosopher, his mathematical physics contributed to the development of new mechanics, technology, and industry out of which the matrix of modern disciplinary society has emerged. Galileo's scientific instruments, or any objects used in some form in his science, not only actualized his new worldview, but also performed the disciplinary function of fastening the body on to the instrument sexualized by the homoerotic rules of scientific inquiry. Galileo's pendulum in many respects illustrates one such moment of contradiction, signifying equally the birth of new subjectivity and the birth of a new disciplinary order.[19]

This is the general point of this book, and it originates from a simple observation. There is a peculiar distinction between our perception of the work of science and that of art. When a musician strikes the right note, we credit the musician's fingers for our delight in hearing it. But when a scientist produces a perfect swing on a pendulum, we see a circumference, the center of a circle; that is to say we credit a theorem, not the scientist's hand. For some reason, when the instrument approaches the ideal mathematical form, the scientist's bodily work vanishes from our sight along with the physical substance of the instrument and we *see* instead only geometry in its place. Is mathematical physics a black hole of the body? Although antithetical, both reactions involve pleasure. While in the first case we have a pleasure of the senses, in the second we have the pleasure of a rational construction.

This peculiar phenomenon is not a natural but rather a sexual one and its explanation may be traced to Descartes's theory of perception.[20] Let me elaborate on this point and introduce Descartes's epistemology in the study of Galileo's pendulum as a case of discourse on sexuality suited for the new form of homosocial domination already outlined in the Neoplatonic way of life. In his optical studies Descartes sought to establish principles of perceptive certainty in the world of sensual illusions by emancipating perception from all external material conditions and basing it on geometrical order. To illustrate his point empirically, Descartes invites the reader to perform a simple but amazing experiment. He wants the reader to find the eye of a dead human or of a larger animal, perhaps an ox, and to peep through it.[21] If the reader does so, Descartes explains, the reader will see objects on the other side of the eye arranged along a perspectival line. The objects are seen by the living eye as out of order, yet reorganized geometrically by the "dead eye." This optical phenomenon led Descartes to claim that the eye, when emancipated from the impressions of life, operates as pure *cogito,* revealing that our perceptions are constituted not empirically but rationally. For Descartes the consequences of this recognition are twofold. First, the eye of the mortified body, blind to all worldly impressions, can invent its own world according to its own formative laws of exactness, precision, and objectivity, which, created from our will and reason, *res cogitans,* removes any doubt about its existence. Second, and most importantly, the rational world, the world of *res cogitans,* assures personal salvation.

What, then, is the relation between the epistemology of the personal salvation of man from woman—of reason from sensual body—and Galileo's pendulum? Although Galileo never formulated epistemology in Descartes's fashion, he nonetheless set the stage with his geometrized pendulum for Descartes's optics to come and, vice versa, Descartes's optics articulated

Galileo's implicit relation to the body, central to his identity as a mathematician-natural philosopher and built into the workings of the pendulum. From the perspective of Descartes's optics, Galileo's pendulum appears to mortify the eye with its geometric order, that is to "emancipate" it from the material substance of the pendulum for the duration of the successful experiment. In the light of Descartes's theory of the "dead eye," this simple phenomenon of Galileo's pendulum establishes an important link between the body, the instrument, and the scientist's "technology of the self." Not only is the instrument here an extension of the body, but also the body's internal constitution seems to be transubstantiated and dematerialized by the instrument in the same way in which Descartes wants to dematerialize himself and become free from the oppressions of nature. This epistemic moment described by Thomas Khun as charged with the "paradigmatic shift,"—for Descartes, a moment of an achieved self-hegemony over the body and nature by pure *cogito*—passes now as an endlessly performable and analyzable body of science blessing us with seeing natural law through Cartesian technique of self-hegemony. The vanishing materiality, the "dead eye," and "pleasure of analysis," though on the surface not related, all merge into the practices of Descartes's and Galileo's subjectivity, and Kuhn's history.

Galileo's attempt to disqualify secondary qualities from knowledge with his mathematization of physics and human experience, appears as Descartes's desire to dissolve the body into pure mind and the senses into mathematical axioms, which is another example of Descartes building a homosocial epistemology. A desire to achieve the ground of absolute certainty and freedom from the oppressions of nature, is for him, both personal and deeply sexual. And, as Jacques Lacan observes, Descartes's path of certainty is "the way of desire . . . desire for certainty."[22] Descartes's first move towards certainty, to bracket all knowledge and all experience, is not a scientific but a "salvational" one, and can be achieved only by an ascetic will.[23] In the postlapsarian world profuse with libertinism, Descartes had, as had many of his famous contemporaries, identified the oppression of nature with feminine sensuality and justified and explained hegemony over the body as masculine self-actualization.[24] He makes this distinction abundantly clear in his letter to Vatier in which he apologizes for having left many points unexplained in his *Discourse on Method,* because: "these thoughts did not seem to me suitable for inclusion in a book which I wish to be intelligible in part even to women while providing matter for thought for the finest mind."[25] Descartes acknowledges that his readership is gendered and that a sexual differentiation carries epistemological hierarchization, male "finest minds" and female, not-so-fine minds, a distinction, upon which he insists throughout all of his work, between sensual and rational perception. Descartes defines feminine perception as *mimetic*. It absorbs and reproduces impressions forced upon the senses; it is epistemologically inferior; it does not reflect itself; it is not aware that it perceives, and it does not mind errors. The feminine mind is, for Descartes, like wax, shaped by pressure and ridden with errors; and because it lacks any autonomy in relation to nature, that is, any power to resist it, woman's "mind" is "weak." In contrast, masculine "mind" is constructivist and "strong" precisely because it is aware that it perceives things, that perception is rational and may be corrected, and that in resisting nature it empowers itself to recreate nature out of its own strength. Unlike mimetic perception, which grows out of a living nature, Descartes's masculine eye has its own pure, internal reason. On the one hand, he Christianizes the masculine eye by making it pure, but on the other hand, he finds in a "pure" eye a source of unlimited pleasure in this masculine mode of perception.

Judging from the above exposition we may claim that Descartes's theory of perception is in part a discourse of pleasure well suited to a homosocial mode of domination. Precisely because in the dichotomized world of the body/woman and cogito/man, Descartes seems to have only one option that will allow him to claim his subjectivity, and that is to turn women into the effect of his self-hegemony. Descartes empowers and sexualizes his position. The mortified and geometrized eye, in this context, sees woman as impure matter and emancipates masculinity from the oppression of femininity. This is, for Descartes, nothing more or less than a starting point from which a total reconstruction of the world, and a creation of new matter may come. To achieve masculine emancipation Descartes, in order to hold the ground of absolute certainty, eradicates all desires toward woman and confronts her in his new self-hegemony as a God who has total power over her precisely because he does not want her.[26] With this in mind, Descartes reaches inner peace, truth, and the infinite pleasure of reason without a woman.

I am discussing Descartes at such length because Descartes articulated clearly homosocial epistemology already situated in the Neoplatonic position on sexuality and rationality governing academies of sciences with which Galileo was associated. When Galileo entered the *Accademia dei Lincei* in Rome and had to swear an oath to chastity requiring him "to avoid 'the attraction of Venus,' 'bad women and profane love,' 'Venereal lust,' 'prostitutes,' 'tempting lust,' 'low passions of the body,' 'carnal drives,' 'libidinous excitements,' and 'the body's innate desires,'"[27] he was sexualizing science even before Descartes articulated it. An organic part of the sexualized body, our eye, one could argue, underwent an ethical and disciplinary transformation through Galilean science. We don't see the materiality of the pendulum and the hand of the scientist in the shadow of geometric law, we may admit now, because we are not allowed to see women, that is, the body-work put into the making and swinging of a pendulum. We can see natural law only if we naturalize Cartesian exclusion. If we follow for the moment Descartes's misogynic distinction between feminine and masculine perception, we may ask ourselves whether we've been *trained* to perceive music in a way Descartes saw as "feminine," and science in a way he saw as "masculine." If so, are we not naturalizing Cartesian misogyny in the course of our observation? In the context of a sexualized history of the senses, we may answer, Galileo's pendulum (not an innocent toy but a sexualized object) delivers truth only by virtue of reproducing a sexual history of its own making. Galileo's pendulum thus swings not only as matter purified from femininity, but also as a signal of the infinite pleasure of masculine domination.

My strategy in this book is two-fold: first, to desexualize Galileo's pendulum by situating it in the history of Western sexuality, and second, to recover and to account for what has been excluded from Galileo's science as the hidden value inherent to the *body-work* of science—its "handmaid."[28]

My first task shall be to desexualize Galileo's pendulum on two fronts. One is the reconstruction of the sexual history of the body.[29] The body: the inscribed surface of erotic events, which Galileo, Descartes, Boyle, Newton, and the Jesuits inherited as a sexual artifact and used both as an instrument in experimental science and as a tool to sexualize their discourse by discrediting the body as the scene of supreme pleasure. The other front focuses on the discourse of scientific rationality itself and the ways it continues this history of sexuality. By relating these two, I hope to demonstrate that the history of western sexuality reveals that the aim of the disciplinary mechanisms of the desexualization of the body is not to eliminate sex, but to emancipate it from the body and elevate pleasure into the sphere of a pure discourse. Thus ignoring this history does not negate its presence in modern science, I would argue, it only proves that in order for modern science to be pure "sex," it does not need the body as its referent.[30]

It may appear to the reader that I am shoring up Freud's argument from *Civilization and Its Discontents* that the civilizational process increases zones of sexuality and erotic pleasures precisely through increased sublimation of sexual desire. The Neoplatonic concept of "objective knowledge" has become enmeshed with the Neoplatonic concept of love, making truth out of "sex" and "sex" into truth. Psychoanalysis would treat seventeenth-century natural philosophy as a mode of homosexual sublimation, which creates new intellectual values. However, my position on sexuality is not psychoanalytic. Like Foucault, I view "sex" as an artifact of discourse on "desire," not as libido with science as its sublimation. My concern is not what seventeenth-century natural philosophers really did with their libido, but rather how present-day history's asexual account of science turns into the discursive proliferation of "sex" through the pleasure of absent pleasure.

Foucault displays an enormous distaste for the concept of "desire" and its normalizing power. To counter it, he has advanced the concept of "bodies and pleasures." Arnold Donaldson makes clear what the difference between these two concepts is for Foucault: "Desire has a psychological depth: desire can be latent or manifest, apparent or hidden; desire can be repressed or sublimated; it calls for decipherment, for interpretation; true desire expresses what one really wants, who one really is, while false desire hides or masks identity, one's true subjectivity."[31] Whereas "pleasure" remains always on the surface "desire" emerges from the bottom up, bringing the subject out of its abyss; whereas "desire" calls for scientific representation, "pleasure," on the other hand, resists its representation and can only be intensified, increased and modified. These two conceptions of sexuality bring about two conceptions of knowledge and two different relationships of knowledge to power—one normalizing, and the other resisting.

Just as "sex" may be employed for the proliferation of normality, so, Foucault theorizes, it may be employed to create ruptures inside the same process. With this in mind, Foucault contrasts a prediscursive sexuality, *ars erotica,* common to the Orient with a discursive *scientia sexualis* developed solely by the West. He equates *ars erotica* with "bodies and pleasures," a sexuality of practical intervention that induces, intensifies, and modifies polyvalent pleasures without ever transforming pleasure into representational practices. *Scientia sexualis,* on the other hand, transforms pleasure through representational practices into the pleasure of absent pleasure, or "desire." Although Foucault later retreated from the concept of *ars erotica,* considering it inapplicable to our time, he has left the concept of *scientia sexualis* ambiguous enough for us to seek ways out of the psychoanalytic normalizations. Just as Nietzsche, reversing Apollonic and Dionysiac principles, calls Apollonic coldness a kind of passion, Foucault ponders if *scientia sexualis* has become *ars erotica* subjugated to the "pleasure of analysis." As Donaldson posits, "pleasure of analysis" is at best an ambiguous concept. However, following the adage "Where there is smoke there is fire," we may certainly claim that where there is "desire" there must be subjugated pleasure. For us, then, the problem becomes how to pervert perversity and reclaim pleasure from the pleasure of absent pleasure.

One may well ask, "How can discursive sexuality be desexualized discursively?" Is not "pleasure of analysis" a starting point for any history of pleasure? Is my position tenable? It is true that to desexualize science would imply a recovery of the body and desexualization of the "pleasure of analysis." This is only one solution; the other is in the way of desexualizing scientific discourse by means of *inventing* its sexualized history. I do not mean writing an *actual* history of this discourse but rather inventing a critical view on it, looking at it as if through the "dead eye" of an "ox" and seeing sexualization of science through the sexualizing eye which created it. Here in plotting my strategy I explore Freud's concept of "repressive sublimation" for heuristic purposes only.³² He defines "repressive sublimation" as the blocking of pleasures from their objects of desire by repressive reality principles in order to achieve social aims such as, for example in our case, homosocial domination over heterosexual drives in order to advance scientific rationality. Genealogically speaking, "repressive sublimation" is not a mental process but a discursive account of specific bodily practices and rules of uniformed conduct. In the context of seventeenth-century experimental science, we may then claim, managing experimental settings and conducting instrumental demonstrations or measurements (requiring disciplinary conduct and homosocial bonding on the part of the seventeenth-century scientists), illustrates the practical foundation of "repressive sublimation." Homoerotic ties among men of science, whether displayed as homoerotic desire among Neoplatonic academies or Jesuit brotherhoods, or heterosexual chastity by the British empiricists, all resort to the body techniques of sexual deprivation and exclusion of women in order to advance a discourse on scientific rationality. The history of science belongs to the same continuum of "repressive sublimation" as long as it does not scrutinize its exclusion of sexuality from its analysis and thus does not recognize itself as a work of "erotic improvisation." Alexander Koyré for one, much like the scientists he represents, continued to negate the body in relation to scientific rationality and in doing so put the sign of his history under the principle of repressive sexuality. In this respect "actual" and "represented" history of science both emerged underneath "repressive sublimation." Consequently, history in the present form, which does not recognize science as an "erotic improvisation," comes to us as a repressive pleasure; a language of history in its disconnect with the body is, to borrow Virginia Woolf's exquisite metaphor about homosocial language, like a fetus aborted from the body and kept in a "glass jar."³³ The strategy of "desublimation," that is, unpacking the genealogy, may not eliminate "sex" from discourse, but if pitted against itself, if the history of science as a product of "repressive sublimation" is subjected to (let me twist Herbert Marcuse's terminology with a dash of irony,) "repressive desublimation," we may, in sexualizing sexualization, desexualize the desexualization of Galilean science. Putting "sex" under its own totalizing and regimenting surveillance, we resist the natural tendency of "sex" to conceal itself behind reason and, by entrapping sex in its own gaze, we may neutralize some of its oppressive force.

This brings us to the second focus of my strategy, the history of science. The history of Galilean science is not a breaking point in the history of sexuality but rather its continuation into the present by means of concealment. Consider for example the twentieth-century history of Galilean studies, the "reconstructivist"³⁴ uprising against Alexandre Koyré's deductivist approach to Galileo, and Koyré's provocative claims that Galileo never performed any experiments and therefore never made any discoveries via experimentation.³⁵

While Koyré claimed that Galileo's science was deductivist, "reconstructivists" argue that Galileo was experimentalist. Although appearing to be at opposite poles, the "reconstructivist" and the apriorist positions maintain a similar attitude towards the body. Koyreans regard the bodily dimension of Galileo's physics as irrelevant and stress pure mathematical reasoning as the basis of his mathematical physics. "Reconstructivists," on the other hand, emphasize the reconstruction and repetition of experiential results in order to discover the existence of a technical coherence in Galileo's experimental procedures, while failing to account for the history of the body that, as "handmaid," makes experiments reconstructable. Because of this omission these perspectives subscribe to the logic of sexualized binaries and their histories reinscribe the subjugation of the body-work in science to the history of sexualized rationality. In their accounts of Galileo's science these perspectives allow the history of sexuality to pass as "Galilean science," rendering them unreflected objects of this history.

Unlike Koyré and the "reconstructivists" I will address Galileo's pendulum not as a relic of a history that illustrates Galileo's deductivist or inductivist science, but rather as a point of departure in locating what is missing from this history and without which this history would not be possible. I intend to account for the knowledge that the body itself produces in relation to the instrument, knowledge that because it is informal and intuitive has remained excluded from the formal knowledge of Galileo's science. As Robert Markley observes, Boyle invented a new style of scientific writing which strategically omitted the actual work of the experimentation in order to create a fictional narrative or "virtual witnessing" of the experiment. "No descriptions of Hook," Markley comments, "or other technicians madly 'playing the Pump' to maintain the semblance of a vacuum are included in Boyle's essay."[36] Hook, "the Jack of all trades" while serving as a "handmaid" to Boyle, remains unrecognized as a part of Boyle's experiment. For Desiree Hellegers it is not a coincidence that those who pushed alchemists and women out of laboratories used a "'progressive' support for scientific discovery" to provide "divine sanction for the oppression of women and the exploitation of the laboring masses."[37] The homoerotization of scientific rationality, whose sexual hierarchies stimulated the proliferation of the seventeenth-century scientific academies, reinvented itself inside the laboratory around the experimental work. Post-Baconian science, in its attempt to desexualize the body of the scientist, resexualized laboratory work, skills, and instruments. "The Jacks of all trades," regardless of their male gender, became new women, or the old "handmaids" to the divinity of homoeroticized science. But precisely this "kidnapping the practice of science," Bruno Latour proclaims, "is what made us modern."[38]

Finally, let me turn to my last point. To achieve that I turn to ethnomethodology. An ethnomethodological approach to science as a method of the *knowing*-body rather than as a discourse allows me to reclaim "kidnapped practices" from that discourse. Officially ethnomethodology studies ordinary practices; it describes and analyzes their prediscursive rationality, the rules we follow in accomplishing our ordinary tasks such as standing in line or, experimenting with Galileo's pendulum, the rules established by the *knowing*-body. For example, if there are three balls hanging on three different lengths of string that must be simultaneously released in order to demonstrate the law of pendular motion as we will see later, then the body must establish the rules regulating the link of the body with the balls required in order for the body to accomplish a rational project of scientific demonstration. The rules are unrecognized by the formal rationality of science and yet only through them is the reality of social action accomplished. Galileo's pendulum comes to us through the history of science either as a sign in a discourse or an artifact. From the ethnomethodological standpoint, Galileo's pendulum is a shorthand for a set of unreflected and prediscursive practices accessible only by means of the body's direct involvement in discovering and incorporating the intelligibility of the body-instrument link. If Galileo's pendulum has, in addition to its material structure, a *performative* component, if swinging it in a particular way, if seeing its path only along certain geometric curves, if more than a pair of eyes and hands must witness and participate in order for something to be a pendulum, then the body cannot access the physical structure of the pendulum in any possible way. There has to be a methodic access into the performative structure of the pendulum, a part missing from its history, which in itself constitutes Galileo's pendulum as it was known by Galileo's *knowing*-body, that is, as the method of inquiry into the body-instrument link.

Therefore a formulated approach to Galileo's pendulum affirms Harold Garfinkel's claim that ethnomethodology is a method of discovering rules and order known only to the knowing body but fended off and disquieting for a scientific discourse of rationality.[39] While science constructs the rules of its method prior to and independently of the unique contingencies of its object of study in order to construct an intelligible order of its inquiry, ethnomethodology has no preconceived theory of the rules of inquiry, prior to and above the *knowing*-body. It is only when the object of analysis is in the body's scope of inquiry that the method emerges in the body as an intuited intelligibility of the object of inquiry. In the way that Sigmund Freud confronted the elusive order of dreams by asking "how much 'method' there is in its nonsense,"[40] so does the body ask "how much method is in this instrument?" But, many will ask, "What is historic about this?" "Is not Galileo's pendulum a fact of history?" "Where is the history in the method?" A succinct answer to these questions is that there is no history in the method of the *knowing*-body precisely because it is the a priori of any history (including the history of Galileo's pendulum). Historicizing the body to the point at which the history of disassociation from discourse becomes transparent may allow the body, by way of ethnomethodology, to step out of its history by reclaiming its status as history's a priori.

I hope to recover the body-pendulum link in its own right, as a constitutive part of the workings of the pendulum, and not as a theoretical model constructed by discourse and known in advance. Without falling back into the naturalized continuum of the discourse, I shall employ my own body (and on some occasions the reader's body) to build and use Galileo's pendulum in such a way as to reproduce a demonstrational phenomenon of mechanical motion as well as to account for the prediscursive labor. In this regard, ethnomethodology may repair the injustice done to body-related practices by the history of science. Here I follow the pioneering works in the ethnomethodological studies of science by Harold Garfinkel, Michael Lynch, and Eric Livingston.[41]

This book is divided in two segments: Part 1, "Pleasure," and Part 2, "Pedagogy." In an attempt to explore the historical modeling of the relation between pleasure and knowledge, in the first part I examine a history of scientific rationality and its relation to the formation of the modern scientist's subjectivity. Here I rely on Foucault's history of sexuality. Since "sex" for Foucault is not a natural pleasure but a discursive force around which the scientific rationality of the Enlightenment burgeoned, I hypothesize that Galileo's pendulum, as an extension of mathematics and the body, must have been sexualized by schemes of historical representation to the same extent that such schemes were rationalized by Galileo. This hypothesis is strategic. I use it to point out the ways in which historians of Galileo's science, by excluding the bodily dimensions of the pendulum, continue to normalize the sexual politics of Neoplatonic asceticism—not as a perverse moment of a bodily resistance, but rather as a discursively normalized perversion—and produce history out of disembodied pleasures. These negative politics towards the body have produced two histories of scientific knowledge: one, a history of formal or prepositional knowledges, which I treat in terms of the pleasures of "discourse"; and the other, a history of "tacit" or suppressed knowledges, developed out of recovering the *pedagogy* of practices as the pleasure of a resistance to a discursive normalization. This section of the book is designed to contest a familiar history of Galileo's science; whereas the second part, *Pedagogy*, explores problems of scientific methodology, and attempts to return the body in an explicit way to scientific practice. Here I rely on the Greek notion of *pedagogy* as the transmission of a precious knowledge from body to body.[42] In contradistinction to a history of Galileo's science that relies on pre-established schemes of representation of Galileo's pendulum, or treats it as a discursive sign, I turn to practical schemes of the pendulum, which involve informal, undiscovered, instructive bodily knowledges, a body-instrument link without which the pendulum could not have worked as a scientific instrument. Here, ideally, the "pleasure of analysis" discovers its genealogy and dissolves itself into a *pedagogy* of practice.

The book progresses from teaching the pendulum as a textual construct to treating it as a bodily practice. In the first chapter, "Pleasure, Knowledge, and Subjectivity," I explore the relationship between science and sexuality outlined above by placing it within a more specific historical context, the history of time. Among other purposes, Galileo's pendulum was used to produce mechanical time, and in this way was the forerunner of the modern clock. Significantly, in the West, technologies of mechanical time originated in part within and because of the Christian project of the desexualization of the body or, to put it another way, out of the desire to construct transcendental subjectivity. Focusing on the parallel between the introduction of mechanical time in monasteries and the discourse on sexuality within Western Christianity, I discuss their mutual influence. The result, on the one hand, was to desexualize the body, but, on the other hand, by the inverse logic of pleasure, to resexualize the very same desexualizing technologies of time. In the second chapter I argue that the history of "objectivity" belongs in part to the history of "perversion," and not only to the history of normalized "virtue" as a model of normality, as some contemporary historians of science claim. An intrinsic relationship between the discourses on sexuality and rationality in Renaissance Neoplatonism allowed Galileo and his fellow scientists to make important discoveries in physics, but these, in turn, were deeply bound to an economy of male homosocial pleasure. Stated briefly, Neoplatonic asceticism required that the pleasure of procreative sexuality be transformed into the pleasure of "objectivity," a structure coproduced through homosocial bonding. To the extent that "objectivity" involves a new sexual pleasure, Galileo's son is not only Kant and his "moral geometry," but also the Marquis de Sade, who in his sexual laboratories and experiments invented mathematical "objectivity" in order to resist, in a mode of perversion, a normalized sexuality. This historical and discursive relationship between perversion and "objectivity" does not cease with Galileo or Sade but rather continues to exist in a more complicated way with Galileo's historians. For example, at the time when Ernst Mach and other empiricists attempted to essentialize the body, to grant it rational experience within the Galilean model of science, Alexandre Koyré, in the fashion of Neoplatonic asceticism, denounced the body and its role in the production of rational truths. Particularly by singling out the hypothetico-deductive model of Galileo's science, he discursively normalized what was the bodily perverse moment of Neoplatonic subjectivity. The third chapter, "The Jesuits' Homosocial Ties and Experiments with Galileo's Pendulum," continues the concerns of the previous chapter, but this time through a specific historical case. I posit, contrary to Peter Dear's discursivist's view,[43] that classical physics and experiment emerged not only from Jesuit mathematical discourse but also, and perhaps more importantly, from their sexual discipline—a method defined by Foucault as a "meticulous control of the body" common to "monasteries, armies and workshops."[44] The Jesuit experiments with Galileo's pendulum reveal the extent to which experimental asceticism correlated with a male homosocial economy of pleasures, which brought with it the "masculinization" of scientific rationality.

Part 2 of the book, "Pedagogy," explains, illustrates, and expands on Foucault's concept of the body-instrument link. Whereas Foucault's early work treats this link as being determined by the rules of discourse, his later texts present the possibility of a specific practical link between the body and the practical intelligibility of the instrument. This is the problem of chapter 4, "'The Body-Instrument Link' and the Prism: A Case Study." Here, drawing upon insights from ethnomethodology, I take the reader through a "respecification exercise" with the prism, an instrument simple enough to provide a praxeological clarification of the body-instrument link. I ask the reader to involve himself or herself practically with an instrument so as to demonstrate how such bodily involvement brings with it a different understanding of the body-instrument link and of subject position. Through this exercise it should also become clear, contrary to Foucault's early view, that the rules governing any body-instrument link are not codes of action but rather a situated intelligibility of practical action. Such a discovery is important because it reveals the limits of a merely discursive treatment of rules, and illustrates how bodily practices establish their discursive parameters. It also sets up the treatment of Galileo's pendulum that follows. In the fifth chapter, "The Formal Structure of Galileo's Pendulum," I explore the theoretical context of Galileo's pendulum, or treat it as a "reading sign," which, given the above paradigm, will prove to be an incomplete and inadequate approach. The final chapter is a "respecification" of Galileo's pendulum in which I report on how to construct and use the pendulum described in Galileo's *Two World Sciences*. This chapter discovers the firsthand informal schemes of the "body-instrument link." As an example of instructive pedagogy, the recovery of such "subjugated knowledge" requires, I will insist, the changing of one's subject position, or the operationalizing of the body-instrument link with the bodily use of the instrument. Only then is it possible to recognize the situated pedagogy excluded from the history of Galileo's science.

PART ONE

Pleasure

ONE

Time, Pleasure, and Knowledge

A bob attached to a string fixed to a permanent point vibrates until, influenced by gravity, it rests at the lowest position—this short description sums up the plain mechanics of Galileo's pendulum. Its material austerity was lush in sexuality as much as in mathematics. In the Neoplatonic worldview, the two do not exclude one another. Because of the fusion between pleasure and abstract rules, Galileo's pendulum resembled the body of a Christian monk. Like the monk's body the pendulum became abstracted substance by discursive rules, both desexualized and offered to the altar of a text. Gilles Deleuze's claim that desexualization "had become in itself the object of sexualization,"[1] makes the two bodies even more alike. Just as the mechanics of monastic discipline in the course of depriving the monk's body of pleasures resexualizes the mechanics of deprivation so the pendulum resexualizes Galileo's mechanics by depriving the scientist's body. Although mechanically ascetic, for some not even a scientific instrument only an object, Galileo's pendulum caused, in the seventeenth century, a "paradigm shift"[2] and, already in the nineteenth century, with its attachment to a clock, provided a temporal matrix for industrial work discipline.

Not only was Galileo's pendulum one of many instruments seventeenth-century science employed in the process of desexualizing the scientist's body, it was also a contributing factor in the overall sexualization of the "society of the spectacle."[3] In this respect I argue that technologies of time, on the one hand, and sexual deprivation of the body of those who advanced technologies of time, on the other, coextend in history, performing an important transition from an economy of pleasure centered in the body to the discourse of rational discipline and mechanical contrivances. The pendular clock illustrates well this transition to the new economy of pleasure. When Dutch scientist Christiaan Huygens mathematically discovered along which part of the pendular arc the pendulum swings equitemporally and attached it to the back of a clock, he significantly improved the precision of measuring time, allowing the clock to be massively domesticated as an instrument of self-regulation. Also, as a consequence of this mechanical improvement, the bodily routine changed dramatically from local and fragmented activities to a stream of synchronized activities, creating a new "automatism of habits." This technical innovation changed forever our immediate perception of time. On the level of patterned sensation, time was no longer noticed as the sound of a bell in the middle of a long silence but rather, Stuart Sherman accentuates, as a persistent mechanical stream of sound, "Tick, tick, tick."[4] Norbert Elias takes note of this change when he observes that, "Timing had been human centered. Galileo's innovatory imagination led him to change the function of the ancient timing device by using it systematically as a gauge not for the flux of social but of natural events."[5] "Physical time' branched off from the ancient and heterogeneous social time into the matrix of a homogeneous standard of measurement. Instead of a concrete experience of time as a compelling force in social conduct—"as one can readily see if one is late for an important appointment"[6]—"time" became highly abstract and measured by uniform mechanical operation. In a word, mechanical time had tremendous consequences for the birth of what Foucault called a "new disciplinary society."[7]

This new concept of time has not only introduced a higher order of discipline but, more importantly, has become a dramatic locus of modern subjectivity. Walter Benjamin describes the change in the conception of time as one from "messianic" to "homogeneous, empty time." This change forced subjectivity, Benjamin argues, to find its "fullness" in the "emptiness" of time, as if to spin itself a web of "Tick, tick, tick." Eric Alliez describes this drama as having a subjectivity "that is not ours but is time itself,"[8] Edward T. Hall, in a more poetic way, sees subjectivity as "complex hierarchies of interlocking rhythms,"[9] and Foucault captures, less poetically, the depersonalizing mechanisms of the production of modern subjectivity through the "silent" induction of temporal schemes of conduct, which, as "automatism of habit" transform "the peasant" into a "soldier."[10] What had been the "time disciplines" arranged around the bell gradually, with the introduction of the pendular clock, now coalesced into what Foucault pronounces a "general method" of order and truth. Similarly Elias emphasizes this subjectification to the temporal grids of conduct as an important civilizational process of involvement through detachment.[11] He writes:

> The conversion of the external compulsion coming from the social institution of time into a pattern of self-constraint embracing the whole life of an individual, is a graphic example of how a civilizing process contributes to forming the social habitus which are an integral part of each individual personality structure.[12]

Mechanical time makes what Jean-Joseph Goux and Alliez call "logic of universal exchange" intersect with Foucault's "automatism of habit."[13]

THE HISTORY OF THE DESEXUALIZATION OF THE BODY AND RESEXUALIZATION OF TIME

Every civilization, according to Elias, rests upon principles of self-constraint. Native Americans, he points out, were trained from early childhood to sustain severe bodily torture without revealing signs of pain. Since every Indian could potentially have been captured by an enemy tribe and exposed to torture, he or she had to be prepared to preserve the tribal pride by detaching from bodily pains and so affirming the civilizational process of tribal society. Such a natural form of asceticism, lying at the foundation of any civilizational process, finds its equivalent in the sexual asceticism of the West, regulated by the mechanical paradigm of time.

Historically, the mechanical measurement of time belongs to the process of involvement through detachment. It came out of monasteries and the development of the "automatic habit" of monastic life. Monastic sexual discipline revolved around a tight schedule of prescribed activities all aiming at some form of detachment from the body that would separate subjectivity from the naturally occurring pleasures of the body and induce the erotics of the discipline: prayer, confession, fasting, etc. In this respect, the development of the mechanics of time is inseparable from the massive project of the Christian Church to desexualize the human body, or more precisely, to establish a disciplinary scheme for the production of a normalized identity. When Galileo used the pendulum as a "time keeper" in his inclined plane experiments or when he demonstrated the isochrony of pendular motion, "time" had been already resexualized and thus could be objectified, mathematized and mechanized. The rule of the disciplinary mechanics that led to the sexualization of "time" engendered the rules of the new mechanics. Galileo's pendulum became a product, and a symptom, of this interlacing.

I am not suggesting that in the pre-Christian world time was not eroticized, only that it was not mechanized and subjected to mathematical discourse. We can read the Greek story of time and realize that although not mechanized, time was eroticized, too. According to Foucault's account, the Greeks sought an internal connection between order and pleasure, and ascribed virtue to the art of proper timing for pleasure. "Timeliness" consisted of determining the opportune time, *kairos,* for the use of pleasure, by means of *askēsis*. "Right timing" for the Greeks had no prescriptive value, but was based on the individual's context and an individual sense of time in relation to self-stylization. Foucault emphasizes the esthetical and ethical value of "timeliness" to the Greeks. In the *Laws,* as he points out, Plato emphasized that the fortunate person was one who could act "at the right time and in the right amount."[14] Failing to determine the "right time" was associated with a lack of integrity and of personal freedom. Those who acted, Plato states, "without knowledge and at the wrong time" would live life without virtue.[15] In certain instances, timing itself sets a norm. According to Xenophon, Socrates claimed that children born of incest would be punished by God not because of the parent-child intercourse, but because "the parents failed to respect the principle of the 'right time' mixing their seed unseasonably, since one of them was necessarily much older than the other; for people to procreate when they were no longer 'in full vigor' was always 'to beget badly.'"[16] There was a clear ethical connection for Socrates between timing and sexual pleasure, whereby timing itself determined ethical substance. "When" pleasure should be used the Greeks determined in their relation to themselves. They distinguished the time of *nomos,* the inner time of action, from the time of *phusis,* or the external time of moving objects.[17] In Aristotle's physics the external time of objects is "the number of motion,"[18] but motion must be marked first by our senses, and by the inner experience of time. Thus Aristotle conceived external time only through and as an extension of the inner time of bodily action. The Greeks conceptualized time, pleasure, and moral order as relational categories, not external to the body but as its internal parameters.

Although inheriting the Greek conceptualization of time in relation to pleasure and order, Christians, because of their negative eschatology, reformed this relation and by doing this they also reformed time and pleasure. Christian time starts with Adam's fall into carnal pleasure and ends with a purification from all carnal pleasures; it therefore has a *telos* that begins with the negative use of pleasure. Because in Christian theology salvation is measurable by means of time, the rational problems of mechanics—of how to have the most precise measurement of time and therefore identify "right time"— held the key to the mechanics of salvation. For example, it has been said that God comes to visit souls in the "right time." To be ready for God's visit is both a moral and a mechanical problem for the Christian; the laws of mechanics and morality necessarily had to merge. "The unfolding of the human race" through time also unfolded two kinds of sexual disciplines, austerity and procreativity, both of which emerged out of the Christian interpretation of time. While it was interpreted that a celibate person made the clock of human time stop in his or her heart by abandoning marriage, allowing the person to be "'on the frontier' of another world," those who opted for marriage and procreation "sought," Peter Brown argues, "frantically to soften the somber tick of the clock of death by . . . [the] begetting of children."[19] In both cases, however, the choice for procreation or abstinence rested on one's interpretation of time. The clock in a monastery or a bedroom, then, was not only symptomatic of modernity, but also of the intricate relation between the conceptualization of time and the regimentation of pleasure.

While Christian eschatology connected time in a general sense with sexuality, the ascetic relation of the Christian monk to his or her body and the sexual disciplines of monastic life laid the groundwork for both the conceptualization of mechanical time and the desexualization of the body itself. Some historians locate the birth of temporal discipline in the church's fundamental ambivalence about emotions aroused by intense mystical experience of the early monks. In their first five centuries, early Christian communities were characterized by ecstatic dancing animated by deep mystical experience, not by orderly and subdued ceremonies. Spontaneous and uncontrolled bodily expressions of religious enthusiasm certainly had contributed to the development of the strong collective effervescence among the Christian congregation, but often, this amorphous mass of mystical ecstasy would erupt into full riots, violence, and even killings, as with Hypatia, one of the famous Neoplatonic philosophers, at Alexandria in 415. Such violence, however, disturbed the Church authorities less than the erotic dimensions of religious enthusiasm. In the fourth century St. Ambrose of Milan mourned the sensuality of exotic dancing, even though he recognized its spiritual function, and recommended that rituals take place in the mind, not in the body. In the next century, St. Augustine expressed his concern about the possible sexual arousal of ecstatic dancing. His ascetic ideals prevailed in the Church and they took the form of the first monastic rules laid in the sixth century by St. Basil and St. Benedict, which set the norm for transforming erotics of ecstasy into ascetic discipline.[20] Historian William H. McNeill writes:

> duly constituted authorities constrained nearly all Christian monks to live together in monasteries and conform to rules, thus ending public outbreaks.... Congregational singing, processionals, and other stately forms of worship played a conspicuous part in the new monastic rules, supplementing most of the private, trance-inducing exercises that individual monks had formerly engaged in. This strengthened public and ecclesiastical order by damping back the unruly emotions associated with a direct and personal encounter with God.[21]

In the context of the de-ecstasization and desexualization of the monk's body, time played an important role in controlling and transforming erotic enthusiasm into a disciplined conduct. Emphasis on discursive rule and its interpretation replaced bodily expression with discourse and bodily discipline. Asceticism in the form of "the Pauline ideal of continuous prayer," historian David S. Landes observes, "in conjunction with ascetic diet" aimed to "promote a state of light-headedness, conducive to enthusiasm and hallucinations, or, euphemistically, to illumination and visions."[22] As enthusiasm shifted from the bodily expressions to disciplined expressions so did the body space became more regulated by time. Similar to the way that pews were introduced in Western Christianity to restrain spontaneous body expressions, isolating one body from another with wooden barriers and reconnecting them through the codes of new public piety, the time schedules of the monastic rules, and later, mechanical clocks, decomposed a bodily sense of time and re-composed and connected bodies through schemes of mechanical time.

The above changes introduced a temporal discipline of the body around which time, before mechanical instruments, was known and measured. Perhaps the central component of this discipline was self-surveillance. The monks' surveillance of their inner predispositions to carnal pleasures had been mediated for centuries by the simple "mechanics of the bell." The sound of the bell reminded monks about the time of prayer, penitence, confession, and any other method of excluding the body from pleasure. The monks' interest in mechanical time comes from improving bodily discipline. Landes, for example, claims that the bell was part of a larger process of depersonalization and deindividualization in closed monastic space, which was collectively and panoptically occupied so that everybody's movement was seen at *all* "times" and as located in *our* "time." Time, in this disciplinary context, was an index of both social discipline and bodily mechanics. Landes concludes:

> there was "only one time, that of the group, that of the community. Time of rest, of prayer, of work, of meditation, of reading: signaled by the bell, measured and kept by the sacristan, excluding individual and autonomous time." Time, in other words, was of the essence because it belonged to the community and to God; and the bells saw to it that this precious, inextensible resource was not wasted.[23]

Similarly, Lewis Mumford characterizes this relation as setting the stage for the future disciplinary order of the industrial society. He observes:

> the monastery was the seat of a regular life, and an instrument for striking the hours at intervals or for reminding the bell-ringer that it was time to strike the bell, was an almost inevitable product of this life. If the mechanical clock did not appear until the cities of the thirteenth century demanded an orderly routine, the habit of order itself and the earnest regulation of time-sequences had become almost second nature in the monastery. . . . So one is not straining the facts when one suggests that the monasteries—at one time there were 40,000 under the Benedictine rule—helped to give human enterprise the regular collective beat and rhythm of the machine; for the clock is not merely a means of keeping track of the hours, but of synchronizing the action of men.[24]

By virtue of the bell and, later, of the mechanical clock, the mechanical production of temporal units became an external scheme of the monk's techniques of sexual austerity essential for the formation of Christian identity. The rules of mechanics became embedded in the monk's practices of subjectivity and also through timely confession into a discourse on sexuality. Christian communities were organized around a rhythmical order, perceived as both objective and cosmic, until eventually bodies were eclipsed by the ever-increasing order of mechanical time. "Canonical hours" were the first disciplinary relation between standardized time and ascetic pleasures.[25] As J. D. North describes them, the hours "were struck eight times daily on a tower bell which, in summoning the monks to prayer by day and by night, was heard far beyond the confines of the cloister."[26] Standardization of the monk's relation to his own pleasures allowed for the *synchronization* of the actions of the others. Benedictine monks, through their daily discipline, gave "human enterprise the regular collective beat and rhythm of the machine," or acted as a human clock that synchronized actions and humans beyond the cloister's walls. The mechanization of the monastic life and its visibility beyond the cloister's walls pressed the idea of order through the temporal uniformity of practice, observance, and enforcement of bodily discipline.

With the introduction of the mechanical clock, monastic discipline—that is, the surveillance and interdiction of carnal pleasure—was strengthened. As the monastic discipline became more instituted around the tenth century with the Cluniac order, the nature of the discipline and the monastic identity changed as well. With the Cluniac order, discipline consisted exclusively of praying. In the eleventh and twelfth centuries, with the Cistercians and particularly the Benedictines, work, in addition to regular praying, also became part of monastic discipline and regulation. As Landes comments, "Discipline in turn had at its center a temporal definition and ordering of the spiritual life: *omnia horis competentibus compleantur*—all things should be taken care of at the proper time."[27] Perhaps the most important consequence of the restoration of monastic discipline and the transformation of the monk's identity came from the invention of "confessional manuals." The extroverted pastoral schedules, as historian Jacques Le Goff observes, "introverted apostolic instruments oriented toward the discovery of internal dispositions to sin and redemption, dispositions rooted in concrete social and professional situations."[28] In confession, "sex" became the most observed sin and at the same time because of it became the code of all bodily pleasures. Foucault comments on this: "A dissemination, then, of procedures of confession, a multiple localization of their constraints, a widening of their domain: a great archive of the pleasures of sex was gradually constituted."[29] "Sex," as a code of pleasure, became a discursive effect of the amalgamation of the mechanization of time and confession. Moreover, as the body became ever more colonized by a moral code, for the monks, night became the new frontier of ascetic pleasure. Forced to practice surveillance over their pleasure, the monks prayed into the night, or practiced "nocturnal offices."[30] They stayed awake and remained vigilant over the flesh, simply by watching the clock making precise time, a task not easy for all to follow.[31] This mechanization of ascetic practices became second nature in monastic life.

The code of sexual conduct, which ensured the purity of the soul, depended ever more on the mechanical clock. Out of this dependency came a new professional knowledge and new pleasures. This was evident in the increasingly public displays of reverence for clocks, which were gradually placed closer to God, as in the house of Austin: "The clock was set up alongside a great painted crucifixion scene, with attendant images of Mary and John, on the rood-screen and loft, or gallery."[32] This dependency demanded that the citizens maintain public clocks, invent a profession of clock-makers, and develop the theory of mechanics. Within the church, timekeeping—which had at first been no more than an aid to the regulation of worship—soon became an "honorable occupation."[33] The two books by the Venerable Bede, a respected scholar of time-measuring, *De temporibus* (On Time) in 703, and *De temporum ratione* (On the Reckoning of Time) in 725, marked the beginning of the discursive control of the skills related to measuring time. While timekeeping once signified the drama of a mechanical cosmos it soon became, with the profession of clock-makers, surrounded by a wide range of more earthly amusements: striking jacks, jousting knights, wheels of fortune—even pornography, as I shall discuss below. The immediate point, however, is that the mechanical clock was from its inception the object of both moral power and ascetic pleasure.[34]

Mechanics served to illustrate and to envision a divine order of disciplined bodies subjugated to the universal order of conduct. A moral philosopher, Dasypodius, wrote that through the making of the clock, people are educated not only in astronomy and mechanics, but also in morality, which should in his view allow them to have pleasure in their contrivance.

> For just as philosophers examine by observing the nature, force, and effects of things, so do mechanicisians bring about with the work of their hands, their industry, talent, and skill those things which are either necessary for life, or made for pleasure, or benefit daily use.[35]

The introduction of a mechanical clock in monasteries, whereby confessions and penitence were standardized, played an important role in the forming of both a Christian identity and a modern social order based on large-scale planning, thinking, and the production of a civilization that tracks time.[36] The clock, according to Mumford, laid the matrix for industrial capitalism:

> The clock, not the steam-machine, is the key machine of the modern industrial age. For every phase of its development the clock is both the

outstanding fact and the typical symbol of the machine: even today no other machine is so ubiquitous. . . . But here was a new kind of power machine, in which the source of power and the transmission were of such a nature as to ensure the even flow of energy throughout the works and to make possible regular production and a standardized product. In its relationship to determinable quantities of energy, to standardization, to automatic action, and finally to its own special product, accurate timing, the clock has been the foremost machine in modern technique: and at each period it has remained in the lead: it marks a perfection toward which other machines aspire.[37]

Not only other machines but also subjectivities and their "imagined communities" have looked to the clocks to see themselves in a new way. Sherman points to the birth of new literary technologies, which emerged as a consequence of the change in the sensual perception of time by the pendular clock. An experience of time as a stream of miniature measured units has changed the way that the self experiences itself through time. Time became the omnipresent matrix around which a new prose genre was invented. "It structured" Sherman writes, "not only diary and newspaper, but also periodical essay, journal-letter, and travel book." The use of the diurnal form enabled authors to write about themselves and see themselves in a new way through the hearing of the "new time," and enabled "readers to recognize, interpret, and inhabit the temporality by which the whole culture was learning to live and work."[38] This new literary technology conveyed a sense of "simultaneity" among the people in time as a means for structuring personal and group identity. Benedict Anderson notes that Benjamin's notion of "homogeneous, empty time" is of fundamental importance for understanding "the obscure genesis of nationalism." "Empty time," Anderson insists, provides a shared matrix of conduct; fostering a sense of "simultaneity" among numerous individuals who have no way of knowing each other and yet may imagine a community with them in time.[39] For Dr. Iwan Bloch, a scholar of sexual science, this simultaneity in conduct or nations produces a patterning of sexuality. He writes,

> Mantegazza certainly is right in asserting that there is a national love, that every people offers something original, peculiar in its sex life. Thus we find a predilection for active and passive flagellation undeniably more widespread among the English than among other nationalities. The great scatological literature of the French points to a remarkable sexual perversion which was already spreading in France in this respect. The same is true of sadism in France. Yet these peculiarities indicate the cumulative effect of purely external factors, like imitation,

seduction, etc., rather than an influence to be explained by national character alone, even if one nationality took up these peculiarities before any other did.[40]

Not only has new time provided the structuring of an identity along the empty schemes of time, but also its empty schemas have gleaned individual sexual practices and "accumulated" them "over the period of time" into a single collective pattern in which a nation acknowledges its own birth as a sexual pattern and in relation to which it organizes its "bio-power."

The invention of personal watches during the Reformation deepened the level of temporal discipline and personalized it even more—it is not a coincidence that it was contrived within the Calvinist milieu—the institutions of the temporal discipline, monasteries, prisons, or factories, were individualized inside a single body. Furthermore, with the invention of the personal watch, timekeeping became not only the profession of some, but part of every profession. That is to say, it played an important role in professional self-discipline. "Necessity," "benefits," and "pleasure," as Dasypodius has argued, were all merged in the mechanics of time. Having a watch was a pleasure of a mechanical order, the pleasure of a work schedule, of a daily ascesis; owning the watch created the possibility of synchronizing the use of pleasure with others, and mechanical time became a new object of pleasure. Calvin saw it as a "marvelous instrument," and "accepted the watch as a useful instrument and thereby enabled the jewelry trade of Geneva to save itself by reconversion."[41] What the clock was for monastic life in terms of group discipline, so was the personal watch for reformed Europeans, for the in-the-world, individualized ascetics of Post-Reformation Europe.

In the eighteenth century, after anticlericalism opened the way for the erotic imagination, personal watches became not only a disciplinary tool but also objects of erotic pleasure. An eroticized personal watch usually showed sexual acts among monks and nuns. As Landes describes:

> These were still paintings, usually crude, but sometimes very well done; the Geneva erotica were the best. In the early nineteenth century, however, the new vogue for automata led to a logical transformation; from stills to animation. The rhythmic oscillation of the balance wheel was peculiarly suited to these simulacra, whose resemblance one to another would indicate that a few specialist suppliers cast the moving figurines. The tiny crudeness, even grotesqueness, of these representations would seem to limit their erogenic value: they are more amusing than arousing. But they must have made marvelous openers for conversation between the sexes, and I suspect that many a Casanova used his secret pocket peep show-these scenes were usually concealed behind special panels or covers-to test his partner's interest and open-mindedness.[42]

Photo 1

The above account lends itself to multiple interpretations: first, that the erotic watches involved a sharp Protestant criticism of the Catholic sexual hypocrisy; second, that discourses on "sex" and "time" are interchangeable. In any event, one observes that with the development of mechanics the symbolic meaning of time transformed into a pure measure of a mechanical operation and through the body-instrument link gradually became a disciplinary scheme of the body. Mechanical time became the measure of a body's movement, a matrix of synchronized actions, a condition of rhythmical gestures. Through time, the body became a scheme of quantity, a choreography of publicly synchronized ascetic conducts. The temporal beats of the mechanical clock had unified monks and nuns into a synchronized and rhythmical communion of sexually sanitized bodies carrying out a vision of a pure community of negative pleasures, or the pleasure of the mechanized ascesis. Thus the erotic watch involved a sharp irony related to the above economy of negative pleasure. Those who renounced sexuality through strict sexual discipline here are represented as having sex and pleasure generated by the mechanical operation of the watch.

This inversion of pleasure from the "sex" of the body to the "sex" of mechanical operation is not gratuitous, but suggests a resexualization of the desexualizing tools and also a transformation of the body. Mechanics shoots through the discourse on sexuality, and its logic produces and measures mechanical time, thereby codifying the body in the same way that "sex" codifies pleasure. "Bodies" and "pleasures" are set aside, codified, and ready for transaction and exchange. The "empty" forms of these two logics of "sex" and mechanics universally exchange here their equivalence; mechanical operation is exchanged for pleasure, and *vice versa*. The body is transformed by this exchange; no longer a stable and fixed natural object enshrined in senses and merged into sensual experiences, it is disassembled and reassembled along the standards of this exchange. This process allows for a "paradox" of body and pleasures created by the interchangeability of mechanics and sex, as discussed by Allucquere Rosanne Stone in her book, *The War of Desire and Technology at the Close of the Mechanical Age*. There she explains how it is possible to have "hetero phone sex" performed by a lesbian. She claims that through both the technology of phone sex and the exchange of senses and dramatic narratives during phone sex, the bodies and "natural preferences" do not restrain the process because technological mediation invents the body in accord to its rules of operation. She writes: "what was being sent back and forth over the wires wasn't just information, it was *bodies*."[43] Prior to "phone sex," the erotic watches had already suggested a shift in the perception about the referent of "sexual" pleasure. The referent, as Stone points out, rests no longer in the "body" but in the *mechanics* of "sex," that is, in the *coded operation* that allows the simulation of pleasure. It is this new referent, one might argue, that divulges "sexual" pleasure as not being fixed by the body but rather as a floating sign that is a universal standard for exchanging bodies for words. The *mechanics* of "sex," it follows, are transferable and interchangeable with the *"sex"* of mechanics. Paul Virilio observes that the contemporary Catholic Church's indignation about "telesexuality" strongly indicates sexual mutation from the erotic watch to the internet; the simulational virtues of high-speed technology, Virilio observes, eclipse bodily union by the flip of a remote, "pulling the plug on the animate being of the Lover."[44] And while the watch, this rudiment of the modern speed technology, could simulate "orgasm" by measuring time, "orgasm," no longer a code of using bodily erogenous zones, appears now in modern speed technologies as a code of electronic transmission.

The nascent *technophilia* of the erotic watch begins to expose *biophilia* as the dogmatics of pleasure.[45] One poet transposes this transferability of "sexuality" into a discourse of machine and pleasure when he says that "The plane is the only thing I've ever really loved" because, like a lover, "The plane tears you away, makes you live dangerously, offers you happiness, brings you back when it's good and ready."[46] Paul Virilio, in similar fashion, comments on this nostalgic love towards the airplane when he writes:

> as with the nozzle on the jet engine of a machine capable of breaking the sound barrier, everything comes together in long-distance love, thanks to the power of ejecting others, to this ability to ward off their immediate proximity, to "get off on" distance and make headway in sensual pleasure the way jet propulsion propels the jet.[47]

Following Galileo's claim that distance is a function of time, one might point out that pleasure is a function of speed. By this logic, instruments of speed must also be pleasure inducing. In becoming speed, the concept of time, which started as a highly contemplative concept enhanced in symbolic meaning and sexual austerity, sinks even more deeply into one's muscles, practices, and pleasures. As a consequence of this transformation, mechanical time has achieved a long-awaited monastic ideal about the eradication of pleasure from the "act of the flesh" and has installed pleasure into the pure code, not only for the religious elite, now, but for everyone. J. G. Ballard's novel, *Crash*, about urban people who "get off" on speed, car crashes, and prosthetics, authenticates this ascetic vision of the body as a machine. Therefore a machine, such as a car, as a sexual force, may after all not be, as Ballard had hoped, only "an extreme metaphor for an extreme situation."[48] As time, through modern mechanics, increasingly became pure speed, the desexualization of the body through the Western history, Ballard claims, is allied with the resexualization of speed technology. Speed, as Virilio argues, becomes "the coitus of the future."[49] While *Crash* may be a bizarre story, it nonetheless implies that a modern technology of time harbors a discourse on sexuality, and allows us to see how mechanics and machines have replaced the ancient discourse on pleasure centered in the body.

The above account may seem irrelevant to historians of science who, while prepared to accept in general the above argument as a wide historical "background," are less prepared to acknowledge a more direct relationship between sexuality and scientific "objectivity." The clock, one may argue, still measures time, and still measures time regardless of whatever design appears on the watch's face. In the next chapter, I will therefore examine how historians of science themselves have not incidentally obscured the role of the body in constructing "objectivity" properly, precisely because their relation to the body defines, to a large extent, their disciplinary relation to the historical body.

TWO

The Perversion of Objectivity and the Objectivity of Perversion

Some historians of science today reject the notion inherited from the philosophy of science that "objectivity" is an achievement of pure rationality and instead argue that it is an achievement of virtue. They trace the meaning of "objectivity" back to various moral, ethical, and aesthetic discourses, all of which belong to practices of subjectivities. Peter Dear points out that seventeenth-century "objectivity" rested entirely on personal "disinterestedness." By drawing on both David Bloor's claim that "objectivity" can be understood only in terms of social history and Theodore Porter's observation that the collapse of scholastic hierarchies gave birth to "disinterestedness," Dear concludes that "objectivity" conceals a history of intricate relations between power and subjectivity.[1] Similarly, Lorraine Daston claims that the history of "objectivity" is a moral history, since aesthetics and moral philosophy, rather than natural science, constitute its native soil. To press this point further, she goes back to the eighteenth-century aesthetics and moral philosophy of Hume, Smith, and Shaftesbury in order to show that what is variously called "contemplative joy," "catholic and universal beauty," and the "virtue of disinterestedness," all, in one way or another, express the conditions of subjectivity, rather than some Kantian apersonal and transcendental rules of reasoning. Beginning with Descartes and particularly with Kant, Daston argues, objectivity's multiplicities of ontological, epistemological, and socially consensual meanings converged, by the second half of the nineteenth century, into a single meaning: "ontological objectivity."[2] Similarly, in her book *Cognition and Eros,* Robin May Schott tethers "objectivity" to asceticism: "the emphasis on distancing thought from sensualism grew out of an ascetic practice by which men thought to transcend the vicissitudes of the phenomenal world, to escape the mortal fate implicit in the natural life cycle of human beings."[3] Steve Shapin's case study of Boyle also treats the subject of the scientist's practices of the self as a gentleman. Shapin demonstrates that Boyle's "life of solitary disengagement," his "solitary retirements," involved not only special mental conditions but, above all, specific conditions of bodily discipline that resembled those of stoic *apatheia*. Boyle, Shapin claims, gains trust for his experiments by virtue of his negative attitude towards his body and pleasures. Boyle's physical frailty, his delicate appearance—he was often described as resembling transparent "Venetian glass"—proved, for his contemporaries, that he was in fact detached from life and was committed to the rational discipline of the civic order. It is this mastery over the body and pleasures that makes him a disinterested observer, but at the same time distinguishes him from the "mere fine gentleman" who, Boyle claims, "was identified as the servant of his animal nature, as unfree in the ostensible taking of his 'pleasure' as any servant acting at the behest of a human master."[4] As Shapin shows, with Boyle, "objectivity" had a distinct bodily appearance and an inscribed asceticism.

The relation between "objectivity" and morality is even more pronounced in James Bono's history of scientific language. He argues that the seventeenth-century language of scientific objectivity emerged as a reaction to the overall moral decline of that time, which brought about an effort among the male-educated elite to reconstruct postlapsarian languages as a way of achieving a moral redemption from fallen nature, nature that was implicated in the fall from grace and the loss of an Adamic knowledge of God. The first step towards this reconstruction was to objectify nature as a woman, stressing the Christian view that woman's sensuality caused the fall and the loss of Adamic knowledge, and to reclaim the lost knowledge by imposing masculine rationality onto nature, thereby re-establishing control over it. As the extension of human decay, languages suffered from reflecting erroneous reality; therefore the role of science was fundamentally moral and redemptive, seeking truth through the moral reconstruction of the self. As Bono writes: "This new Pentecostal narrative of man's fall and redemption hence pointed toward emphasis upon human industry and the observation of nature in all its diversity—and away from a symbolic ordering of nature—that was to become a hallmark of the new science of Bacon, Galileo, Mersenne, Descartes, and Boyle in the seventeenth century."[5] The seventeenth-century scientists, or to be precise, the natural philosophers, by regarding experimentation and scientific instrumentation as a moral prosthetic for correcting the weakness of fallen human sense, aimed to redeem the language.

While it is difficult to take issue with Bono's claim that objectivity is associated with human redemption, I question his desire, along with that of other historians, to tether the origin of "objectivity" to discourses of "virtue" alone.[6] If Bono is right that the language of experimental science is aimed at moral restoration and the reconstruction of the self along the idea of the fall, then, it seems to me, this language falls under the sign of "sex." A discourse on sexuality, in other words, lies under the self's reconstruction. My abbreviated criticism of the above histories, then, is that the history of "objectivity" and "scientific revolution" is not only a history of "virtue" but also of perversion. I am not the first to suggest this angle of interpretation. Margaret C. Jacob, for one, stresses the role of pornography in formulating the "new mechanics" of the seventeenth century. She states: "pornography . . . has never been imagined as relevant. Yet pornography may have more to tell us about the world that gave rise to the new push-pull metaphysics of bodies, both animate and inanimate, than might have previously been suspected."[7] Neoplatonism has, after all, been credited by historians for both the birth of scientific objectivity and the birth of pornography. Paula Findlen, in support of this latter view, observes: "The erotic and obscene literature of the Renaissance developed an elaborate visual currency integral to the formation of pornography." These techniques, she maintains, were not limited to pornography "but permeated the humanist discourse of the arts." Renaissance Platonism in particular, she emphasizes, "recuperated the erotic" in part because it developed a "voyeuristic gaze."[8] In her "Introduction" to *I modi,* the first erotic book of the Italian Renaissance, Lynne Lawner, like Findlen, links Renaissance erotic literature to a Neoplatonic revival of Greek geometry. She observes that erotic images were "rendered with an equilibrium" and with "a fine grasp of geometry."[9] This new representational strategy of the voyeuristic gaze made the secrets of pleasure explicit and analyzable. This strategy would over time develop into what Baudrillard today calls "a dizziness born of the loss of the scene and the irruption *[sic]* of the obscene."[10] While Findlen identifies Neoplatonic aesthetics as being responsible for the birth of hypersensuality in the Renaissance, Jacob stresses the overlapping similarities between the logical representation of the new mechanics and that of pornography, both of which, in her view, mark the beginning of the materialistic representation of the world. She writes:

> I shall also argue here that pornography existed within, and, in some sense described, new urban social networks made up of many men and some women whose lives were privatized, atomized and individuated, rather like the eroticized bodies found in the literature. As will become evident, philosophical materialism was appropriate in the world of, as well as the texts of, pornography. Only the feared metaphysical underside of the new science could explain the ceaseless desire, the random excess, the sheer exuberance of bodies released from traditional moorings and pious inhibitions now rendered irrelevant by markets and presses, now encouraged by pens made all the more active and virile by their anonyminity.[11]

In her argument, "philosophical materialism was appropriate in the world of, as well as the texts of, pornography"; therefore "objectivity" thenceforth could not be isolated from the general context of "sex" and representation. The piety of the seventeenth-century mechanics, Jacob suggests, brought with it new literary technologies of representing nature as a machine which, when applied to "bodies and pleasures," amounted to "the shared exuberance of bodies released from traditional moorings and pious inhibitions." Here Jacob ties scientific piety as a subjective base of "objectivity" to pornography as its oppositional agency; while the former creates a mechanistic representation of nature, the latter breaks "pious inhibitions" in order to catapult the "body," turned by the new mechanics into a pure sign, up into the stratospheric erotics of representation.[12]

How do the above texts, stressing the role of "sex" in scientific experimentation, affect the groundbreaking arguments of Shapin and Schaffer about the role of the English gentleman in developing the protocols of experimentation? First, it is important to note that both are writing histories of subjectivity. "Objectivity" was supposed to rest on the conventions of transcendental rationality, but their historical analysis has demonstrated that these idealized conventions have specific class locations within the moral conventions of a society. As they themselves attest, "We have begun to develop the idea that experimental knowledge production rested on a set of *conventions* for gathering matters of fact and for handling their explications."[13] Thus for them to answer the question of how experiments produce scientific discoveries, they have to claim that the experiment, as a "form of life" founded upon a set of moral conventions, reproduces the normal structures of moral solidarity. With this explanation, Shapin and Schaffer form a triangle of virtue, trust, and truth, suggesting that scientific truth comes through the mechanism of moral normalization, or restoration. Second, however, it is equally significant that "virtue" does not exclude sex, but in fact depends on it. Take for example, the case of Boyle's apathy. If we examine the moral value of "apathy," which Boyle exercised quite intensely and believed to be necessary for projecting social trust, we see it falls simultaneously under the signs of virtue and perversion. Elizabeth Potter explores this issue in her recent book on Boyle. According to Boyle, God required that men performing experiments be chaste: "to construct a new science which required that facts be produced through experiments properly conducted and attested," the new man of science must be "a chaste, modest heterosexual who desires yet eschews a sexually dangerous yet chaste and modest woman." Women "must be chaste to keep men chaste," those women who do not help men in their divine mission, Boyle pronounced "whores."[14] Women must somehow ensure the chastity of the male homosocial order, thus woman's sexuality was productive of Boyle's "apathy." In other words, the exclusion of sex from scientific rationality produces the stability of apathy; that is, of the scientist's subjectivity, essential also for the production of trust among the male scientists. By excluding women from moral virtues, Boyle's apathy produces, by means of opposites, a perverse sexual periphery around his laboratory and thus sexualizes not only apathy and his subjectivity, but more importantly the laboratory itself.[15]

While for Boyle apathy was an ethical *choice,* for Kant it was an *axiom* of "moral geometry"—that is to say a "necessary presupposition of virtue."[16] For both Kant and Boyle, to maintain apathy meant to resist sensual stimuli and natural drives of all kinds, and to replace the natural determination of conduct with a moral geometry grounded in the rules of transcendental rationality. With these same rules Kant resists natural causalities and bestial drives, and hopes to unify the sciences, their methods, and their experiences under the command of a free subjectivity. Thus Kant's epistemology as an ethical event is an *opus contra naturam.* To this *opus* apathy is central, because it subjects transcendental rationality to a subjective will and places the scientist's identity under the power of Kantian "moral geometry." To the extent that a scientist is made of an "ethical substance," a science is a kind of moral *experiment.* I assume that Shapin and Schaffer would concur with this claim. But we differ in our emphasis on the "normal." I take the view that the objectivity of the emerging experimental science relied on conventions for gathering facts aimed not only to normalize, but also to pervert prevailing social norms, as a means of producing new knowledge. We can see the perverse character of "objectivity" clearly by juxtaposing Boyle's account of apathy with that of the Marquis de Sade. Unlike Boyle and Kant, Sade, not ashamed to admit that apathy gives him a great deal of pleasure, states, "The soul assumes a kind of apathy which is soon metamorphosed into pleasures a thousand times more exquisite than those which weakness and self-indulgence would procure for them."[17] Apathy, Sade continues, also creates a new moment of power in one's liberated will that shakes the very fiber of one's being: "the supreme enjoyment of the self will transport him a sovereign, beyond all imaginable limits."[18] This Enlightenment ideal of self-sovereignty marks also for Sade a perverse moment of pleasure.[19] In this scenario, apathy functions not only as a resistance to the naturalized pleasures of procreative sexuality, or to what Foucault calls "the lyricism of orgasm" involving "weakness and self-indulgence," but also as a means to a new de-normalizing pleasure a "thousand times more exquisite."[20] Sade's apathy may be read as a criticism of Boyle, in that by sexualizing apathy Sade brings to the surface perverse revolutionary pleasures concealed beneath Boyle's pietism. Furthermore, like Boyle, Sade finds apathy to be an important method in his sexual laboratory. While Boyle supposedly opposed natural drives in order to find the truth of nature, Sade opposed codes of sexuality in order to find truths about modern subjectivity.[21] The negativity of Sade's perverse pleasures operates here as a prerequisite for the birth of modern subjectivity. Like Boyle's or Kant's apathy Sade's texts negate nature as a substance or a desire. Kant established his "moral geometry" as a weapon of the Enlightenment against naturalism, and when he established objectivity as a scheme of a free subjectivity, he also created a new ethical normality now in a need of negation. Sade, on the other hand, in a method that reflects Kant (albeit with an outcome that contradicts his), performs a civilizational experiment by subjecting Kantianism to the rules of Kantian "moral geometry," and as result gets pure perversion. The Sadean bedroom, as Jacques Lacan points out, therefore appears as "equal to those places from which the schools of ancient philosophy took their name: Academy, Lyceum, Stoa."[22]

Tasking upon Lacan's suggestion in the following section I will argue that historians of Galileo should treat Neoplatonism not only as a metaphysical apparatus that influenced his science but also as a "discourse on sexuality," which ties his rational truths and objectivity not only to a dominant virtue but also, in its resistance to naturalized sexuality, to a dominant perversion. My point will be, then, that it is not the first tie but the second that characterizes the rupturing aspect of Neoplatonism.

KOYRÉ, NEOPLATONISM, AND THE HOMOEROTIC RATIONALITY OF GALILEO'S PENDULUM

Alexander Koyré, the most important Galilean scholar, whose influence in the history of science reaches as far as Kuhn and Foucault,[23] outlines a theory of objectivity which, in his view, begins with Galileo's acceptance of Platonism over Aristotelianism. While making such a revolutionary claim to knowledge, Koyré also (in ignoring the history of the body in relation to "scientific revolution") normalizes a Neoplatonic politic of sexuality. Galileo, as it is well known, admired Copernican science as "a rape upon the senses"[24] and yet a sexual link is visibly absent in Alexandre Koyré's account of Galileo's "scientific revolution." But his omission reads also as the symptom of a particular politic of shame toward sexuality.[25] Insisting on Galileo's Neoplatonism, Koyré reduces Neoplatonism to a pure system of thoughts by excluding the sexual practices involved in its self-fashioning. If mathematics belongs to a unitary Neoplatonic worldview, so do, as we will see, homoerotics and the exclusion of women from scientific rationality. By focusing exclusively on the discourse of mathematics Koyré treats Neoplatonism "Neoplatonically" and to this extent he *naturalizes* Neoplatonic sexual politics, a move that I treat as problematic.

In hitching the birth of "objectivity" to Platonic metaphysics, Koyré makes the famous claim that true empirical science did not begin with the senses cherished by the Aristotelians, but with the application of mathematical axioms. It would be incorrect, however, to leave the impression that Koyré denied the importance of experience in empirical science. He only reworked the definition of an "experience," proposing that it is not a property of the body but the conceptualized effect of mathematical axioms. Thus it comes to us not as a natural but rather a constructed effect of reasoning. He further argues that in order to cast knowledge in a new light and to clarify the logical status of experience in his nascent science, Galileo had to dispute the "natural" version of knowledge centered in the body and senses. The revolutionary achievement of Galileo's experimental science, Koyré claims, is that it disentangles the relation between the senses and knowledge. Because experience and experiment are not complementary in Galileo's science, they stem from two different logical orders. Following Descartes's dualism, Koyré maintained that one logical order is that of a phenomenal body and sense experience, and the other is that of mental constructions. Geometric figures, employed by Galileo in physics, are not empirical objects but artifacts constructed in our imagination.[26] Therefore, for Koyré, theories, mathematical theorems, and mental experiments remain far more important for the formation of "classical" physics and, consequently, for modern physics, than observation, experience, or induction. As Koyré provocatively concludes, "Good physics is done a priori."

> Thus *necesse* determines *esse*. Good physics is made a priori. Theory precedes fact. Experience is useless because before any experience we are already in possession of the knowledge we are seeking for. Fundamental laws of motion (and of rest), laws that determine the spatio-temporal behavior of material bodies, are laws of a mathematical nature. Of the same nature as those which govern relations and laws of figures and of numbers. We find and discover them not in Nature, but in ourselves, in our mind, in our memory, as Plato long ago has taught us.
>
> And it is *therefore* that, as Galileo proclaims it to the greatest dismay of the Aristotelians, we are able to give to propositions which describe the "symptoms" of motion strictly and purely mathematical proofs, to develop the language of natural science, to question Nature by mathematically conducted experiments, and to read the great book of Nature which is "written in geometrical characters."[27]

Here Koyré makes an important epistemological step forward in relation to the sensualistic version of science. He cuts the old bellycord between the body and knowledge, declaring that "experience is useless" in the formation of science, that knowledge doesn't come from the body but is instead constructed. By escorting the body of the scientist out of the theater of knowledge, by denying to the senses and experience the ground of epistemological certainty, by centering knowledge in pure discourse, by empowering the discourse to regiment the world, Koyré—much like the Neoplatonists—elevates mathematical reasoning into a method of truth and grants to it the status of schemes of experience. Axioms, postulates, theorems, and laws are not only epistemic constructs, but in their application, it implicitly follows, they are modalities of epistemological suppression of the sensual experience. Koyré's insistence on hypothetico-deductive and a priori reasoning stresses not only that these forms of reasoning are internal, abstract, and superior to experience, but also moments of power over nature whereby it becomes regimented, interdicted, and subjugated to the mathematical imagination.[28]

Although "power" is not Koyré's explicit concern, it is, for readers of Foucault, difficult to ignore this issue in Koyré's writing. When he represents Galileo's pendulum, Koyré persistently stresses the superiority of pure mathematical reason over sensual experience. Not surprisingly, then, Koyré represents Galileo's pendulum only as a discursive sign or as a "mathematical instrument," concealing its bodily dimension. To achieve the epistemological goal of disembodying the pendulum and separating rationality from its mass, Koyré desexualizes the mass. The pendulum, Koyré declares, is—at least in theory—the "empirically" ideal instrument. By this he means that the cleverly designed instrument virtually eliminates any air resistance or friction and produces perfectly identical oscillations;[29] it moves, in other words, as though it were in the mind. He praises Galileo's pendulum because its minimal material structure achieves two objectives: first, the pendulum eliminates the role of sense-experience, and second, its geometric scheme of circular motion immediately and directly subjugates the senses to its axiomatic structure of a circle. In this case, the scientist *sees* the motion by virtue of mathematical abstraction rather than by a sensual experience. Furthermore, the absence of the sensual presumably allows for the discovery of the formal regularity of pendular motion—in other words, for theorizing.

Out of this "pleasure of analysis" the dominant theory of Galileo's pendulum emerged. The theory states that each pendulum has its own unique time, which is determined by its length. Because of this regularity, pendular motion was the closest to a perfect measurement of time. More importantly, a primitive theory could be attached to the pendulum's order: the pendulum's time is determined only by the length of the string; the time of the pendulum is in an inverse ratio with its length; and the number of swings increases with the decrease of the pendulum's length and vice-versa. According to this theory, it follows that by controlling the length of the pendulum, one could standardize a desirable time of oscillation—preferably the length of one second, soon to become a standard scientific sign of representing motion as well as human labor. Because of its simple mechanical structure, its minimal material content, and its geometrical form (a fixed point of rest, the center of a circle, a string, a circle's radius, a small bob, a point on circumference, a swing, an arch of a circumference) the pendulum was considered kinematically, and thus approximated visually its outward character. The pendular mechanical features, in other words, coincided with the features of a circle, making the pendulum a visual spectacle of embodied geometry. Koyré emphasizes that the greatest achievements of Galileo's pendulum and the source of the pendulum's status as spectacle were its isolation of gravity from the complex of natural forces as a single variable, and its utilization of geometry in observing gravity. The outward principles of pendular motion trained the eye to see how gravity forces the mass of the pendular ball into graceful submission to the laws of Euclidean geometry. Because the axiomatic features of motion were now conceptualized, the center of power for knowledge shifted from the body to reason.

After assessing Koyré's theory of "objectivity" it becomes clear that the history of "body and pleasures" plays no role in it. In spite of this significant absence there is, however, a way to read Koyré's theory of "objectivity" and the history of Galileo's pendulum in terms of the untold history of the relation between pleasure and rationality. Koyré fails to acknowledge that two world systems pitted against each other, Aristotelianism and Platonism, consisted not only of conceptual apparatuses, but also of a specific relation to the "body and pleasures," which is not an external variable but the only means for generating conceptual systems. To make this point clear, we should consider the history of relation between pleasure and knowledge in Aristotelian and Platonic worldviews and then its modality in Renaissance Neoplatonic literary circles.

While for Aristotle, the sensual world initiates knowledge, Plato locates its origin in desire, that is to say in the pure form of *Eros*. Diotima teaches Socrates that the man who sees "heavenly beauty face to face . . . self-unsullied, unalloyed, and freed from the mortal taint that haunts the frailed loveliness of flesh and blood . . ." should be envied.[30] Unlike Aristotle's sense-based truth, Platonism grounds truth outside the sensual world, which forces a philosopher to exercise asceticism, both as a prerequisite for seeking truth and as a resistance to the world of illusions.[31] Consistent with Plato's notion of knowledge as pure desire, this ascetic position is not pleasureless. It merely lacks sensuality. Since truth is a pleasure of pure form, according to Plato's theory of love, mathematics is the closest to pure *Eros*. Consequently, according to Plato's epistemology, mathematical axioms are simultaneously the ground of certainty and of sensual pleasure. In this context, then, any debate over Aristotelian physics, grounded in sensual experience, versus Galileo's mathematical physics, grounded in mathematical axioms, discloses itself as a struggle over the body and pleasures as a *method* of truth.

Unlike Koyré, the Greeks saw the relation between pleasure and knowledge as quite logical. As Foucault points out in the *Introduction to the History of Sexuality,* Greek *askēsis,* a method by which a philosopher produces truth by shaping his own pleasures, mediated pleasure and rationality. This remains true for mathematics as well. Both in *Eudemian Ethics*[32] and *Nicomachean Ethics,*[33] Aristotle, Foucault emphasizes, outlines three prerequisites for successful philosophizing: *phusis, mathēsis* and *askēsis.*[34] Here Aristotle is typical of the Greeks who insisted that in forming a "free citizen," "*mathēsis* alone was not sufficient; it had to be backed up by training, an *askēsis.*[35] (Socrates was very vocal on this point. In his teachings he claimed that philo-sophia begins with one's control of Eros through *askēsis.*) Following Aristotle, subsequently, pleasure is subject to *sōphrosynē* and *enkrateia,* a state of wisdom and virtue resulting from one's mastery over one's possible enslavement by mere pleasure. Pythagorean and Platonic academies were known to require *askēsis* as a prerequisite for gaining mathematical knowledge.[36]

If Foucault's thesis is even partially correct, the emergence of Greek science coincides with Greek *askēsis.*[37] In this respect both Platonism and Aristotelianism belong to a history of pleasure although they each involve a different use of pleasure, or different aesthetics of truth. Aristotle links pleasure to sensation, Plato to pure form. For Aristotle, there is "a direct proportion between the intensity of pleasure and the knowledge derived from sensation,"[38] so the pleasure of sensation initiates a "natural" desire for knowledge. Aristotle assumes that rational truth can be achieved without ever breaking the link with the pleasure of sensation. Foucault elaborates on Aristotle's sensual/conceptual relation: "The desire for knowledge, given at the beginning of the *Metaphysics* as universal and natural, is based on the initial adherence already manifested by sensation: and it assures a smooth passage from this first type of knowledge to the ultimate knowledge that is formulated in philosophy."[39] For Plato, in contrast, the pleasure of sensation leads to error, and truth, for him, is in the pure form that, like a flame in the cave, can be perceived when reason is freed from the sensations of the body. Neoplatonists, particularly Christian Neoplatonists starting with Augustine, later radicalized this position by totally denouncing the body. While Aristotelian *askēsis* cultivates the bodily senses as a vehicle to truth, Platonic *askēsis* cultivates its *negation* as a vehicle for truth.[40] From this perspective, Galileo's science does not represent only the change of a dominant metaphysical system, as Koyré insists, but also a history of the "body and pleasure."

For the Neoplatonists this relation was neither ignored nor questioned. Consider, for example, Cambridge Neoplatonist Henry More, condemning an alchemist who hovers over nature as over the naked body of his bride: "Thou has not laid Madam Nature so naked as thou supposes, only thou hast, I am afraid, dreamt uncleanly, and so hast polluted so many sheets of paper with thy

Nocturnal Conundrums *[sic]*."⁴¹ But one should not presume that Neoplatonism means sexless science, but rather a sexualized science of a different kind. When a Neoplatonist repudiates the alchemical sexual metaphor of science as a *procreation* between the scientist (males) and the nature (female), it is because he sees procreation as representing a corrupted rationality by the body. For the Neoplatonist, the *pure* rationality of Platonic Eros, can be achieved only through homoerotic desires. This desire accomplishes two important things for this worldview: first, it gives an *aesthetical dimension* to Neoplatonic knowledge in claiming truth only through the practices of *beautification;* second, as Herbert Marcuse pointed out, homoerotic desire represented a "protest against the repressive order of procreative sexuality."⁴²

Finding pleasures in rationality and homoerotic resistance to the procreative sexuality of the Catholic Church, had been inspired by the introduction of Neoplatonism to Renaissance Italy. When Marsilio Ficinio (1433–1499) introduced Plato to the Catholic world by translating his works from Greek to Latin, he not only popularized Plato but also, as Giovanni Dall'Orto attests, advocated Platonic love as congruent with the Christian sexual politics of chastity.⁴³ Here we see that the legitimation of Neoplatonism begins with neither political nor epistemic, but with sexual legitimation. Ficinio pressed this point by revealing Plato's theory of love, *amor Socraticus,* as true Christian love between chaste men. "Amor Socraticus," Dall'Orto explains, "was a very philosophical, highly stylized, and rather mystical way of interacting between two men. To work properly, it required of both partners the will to go beyond; that is, to use the other's beauty and knowledge, to God."⁴⁴ Even though Ficinio presented Plato's theory of love "as a delicate, refined form of relationship among highly cultivated individuals,"⁴⁵ Aristotelian theologians remained unimpressed, suspecting brewing pederasty. George of Trebizond, for instance, a fifteenth-century Byzantine Aristotelian, accused Plato in his comparative studies in 1455 of sodomy, and condemned his philosophy as "subversive of Christian honesty, charity, and orthodoxy."⁴⁶ While Ficinio sought Platonic love as an ideal relation common only to monastic settings or to Neoplatonic academies, George condemned it in terms of its sexual consequences. Although apparently unrelated to the issues of science and to the question of whether or not mathematics can be applied to physics, this debate among the fifteenth-century Italian humanistic intelligentsia over the question of love is relevant for knowledge and for a means of conceptualizing it differently.

Ficinio's introductions of the Platonic erotic into Renaissance Italy took many directions. In addition to science and visual art, Italian Renaissance erotic literature organized its narrative around the relation between Neoplatonic homoerotic pleasures and knowledge. Sometimes supportive and sometimes critical, this literature postulated the tie between sex and learning by arguing that "secret learning and secret pleasures intertwined."[47] Sex was treated in Italian erotic literature as a *method* of truth and the tie between pleasure and knowledge used as an occasion to mount a cultural criticism of its time. The erotic literature falling under this category was not scarce. Findlen argues:

> In sixteenth-century Siena, members of the *Accademia degli Intronati* produced such works as Antonio Vignali's *La Cazzaria* (1525–1526) and Alesandro Piccolomini's *La Raffaella* (1539). In seventeenth-century Veneto, the *Accademia degli Incogniti* had among its ranks Francesco Pona, Ferrante Pallavicino and Antonio Rocco, authors of *La Lucerna* (1625), *La Rettorica della puttana* (1642) and *L'Alcibiade fanciullo a scola* (1652) respectively.[48]

In using genitals, intercourse, and pleasures as metaphors of power, such books reflected the class-based distribution of knowledge and pleasure. Although generalizations always fail to represent the heterogeneous fields of human sexuality, it is fair to say that in Renaissance Italy, while aristocratic and educated men preferred Neoplatonic education, which may have included anal sex, lay people preferred heterosexual intercourse.[49] This division, at least in Renaissance Italy, stimulated social criticism through pornography. For example, in *Sonnetti lussuriosi* Aretino portrays sodomy as important for powerful men: "Men insist upon anal intercourse in several sonnets, crying out that 'He who is not a buggerer, is not a man'; and 'May my lineage die out with me.'"[50] Here Aretino makes a parody of the homosexual practices that were preferred by the educated nobility and were affirmed by writers such as Vignali. On the other hand, Nino Borsellino observes that works that affirm homosexual practices, like *La Cazzaria,* should be seen also as "a critique of humanist practices"[51] because they represent the textual gaze of the noble humanist as an erotic one. In *La Cazzaria* Arsiccio (the author) initiates a dialogue with Sodo, a young academician, by pressing the point that sexual knowledge should be the core of natural philosophy. Someone who aspires to "knowledge of natural things must" he insists, "be versed in 'the most natural things' to claim any philosophical competence." Sodo resists on the grounds of Neoplatonic purity, saying: "[M]y philosophy does not deal with pricks or asses, and I am not ashamed to not know them since my studies are founded upon neither the ass nor the cunt, but in more perfect things. . . ." Not satisfied by this answer Arsiccio presses further: "[M]aking yourself touch your prick with your hand is one of the first things one should learn in philosophy. . . . Then, from the mixing of the cunt, prick and ass, knowledge of fucking and buggery follows, and thus scientia is enlarged." After his pictorial articulation of sexuality as a method of inductive reasoning, he then introduces, according to a well-plotted plan, an Aristotelian argument for homosexual intercourse by stating that if different anatomical parts fit each other, then nonprocreative sexual practices should be justified on the basis of this argument. Findlen tells us: "Aroused by Arsiccio's logic *('Arsiccio, tu mi tocchi I lombi con questo tuo ragionare')* the conversation moves to a bedroom where, in the tradition of Sade, pornography and philosophy happily intermingle."[52] The sexuality of rationality and the erotics of education transpired in the erotic literature of the Italian Renaissance.

In counterreformational Italy, the Catholic Church had strengthened sexual codes of procreative sexuality and intensified prosecution of homosexuality. Under such hostile conditions of sexual regimentation, as Leonard Barkan has argued, Neoplatonism developed an entire literature on the sublimation of forbidden sexual practices. He posits that the centuries-long conflict between hermeneutical interpretations of the Old and New Testaments, of Christianized paganism, and the long-standing gap between the erotic and the transcendental were resolved by the Neoplatonic homoerotic economy of pleasure. "Orpheus, the very exemplar of the occult," is also, Barkan attests, "the *author* of pederasty—that is, its originator and its justifying authority. . . . It is an epistemology of inversion, . . . and anathematized forms of pleasure, while still rigorously excluded from the permissible in daily life, turned into the most powerful tropes."[53] Sublimating secret sexual practices into rationality gave rise not only to what is loosely labeled the "scientific revolution" and the new center of power/knowledge, but also to the mechanisms of its resistance, a "desublimation" by erotic literature. In *La Cazzaria,* Vignali claims that the teaching of grammar will lead to the practice of sodomy. Echoing the membership rules of *Accademia dei Lincei* faced by Galileo when he moved to Rome, Filotimo warns Alcibiades that he must separate the affairs of men from those of women because to spend too much time with a woman is to "lose oneself." Admittedly, the sexual identity of the Neoplatonic humanist is ardently masculinist. In any event, in the rhetoric of Italian Renaissance erotic literature, knowledge comes, as Rocco plainly puts it, "thrusting . . . up the ass."[54] The path to erudition in *amor Socraticus* demands a friendship that accompanies intellectual affinities through a sexual transfer of "virtue" from master to pupil producing "perfectly learned" scholars. Most importantly, the emulation of Platonic science, assumed to begin with the sexual imitation of the ancient Greeks, has been, in its sexual idealism, nurtured by the fine arts, but a resistance to the Greek sexual idealism has been promoted through pornography. Although a paradox of sorts, Neoplatonic humanism remained consistent with its principle claim about the relation of pleasure and knowledge, reminding us, Findlen concludes, "that the alignment of politics and learning that shaped Renaissance society could not be divested from its sexual practices."[55]

Renaissance visual art in Italy was no less influenced by the erotics of Neoplatonism than was the literature. The visual art of the Greeks has long been recognized as erotically arousing; Church fathers such as Clement of Alexandria warned Christians about its sexual powers, its ability to create temptations and illusions. Neoplatonism, one should argue, provided not only an alternative epistemology to Aristotelianism but also an alternative sexuality. The epistemic and the erotic merged together into a single etiquette of the Renaissance court. For example, in *Book of the Courtier,* Castiglione discusses the "unchaste desire" of the senses: "If the soul allows itself to be guided by the judgment of the senses, it falls into very grave errors and judges that the body in which this beauty is seen is the chief cause thereof . . . and hence, in order to enjoy that beauty, it deems it necessary to join itself as closely to that body as it can. . . ."[56] Just as our body misleads us in determining itself as the cause of the free fall, so it misleads us in discerning the true beauty; truth and beauty, rationality and pleasure, all stand in transcendental relation to the body and can be seen only through the voyeurism of the "dead eye."

From the above analysis it should be clear by now that Italian Neoplatonism played an important role in redefining the concepts of "experience" and "pleasure" by tying their definitions to the discourse on truth and beauty rather than to nature. In this regard, Neoplatonism was responsible for defining both sexual and scientific experience, but historians like Koyré subjugated the former to the latter. Similarly, Peter Dear, in his latest book on the Jesuits, discusses the construction of the seventeenth-century scientific "experience" while Dominique Dubarle insists that Galileo's science differs from Aristotle's precisely because the latter relies on authority while the former relies on "collaborative experience."[57] Dubarle's point introduces the notion of intersubjectivity in addition to the constructed character of a scientific experience.

Indeed, Galileo referred constantly in his letters and books to collaboration, instructing others to use friends to *perform* experiments *collaboratively*.[58] Maybe Dubarle is correct—so long as one understands that Galileo's experience is not natural and personal, but constructed by a "collective experience"[59] of witnessing and, as Schaffer and Shapin argue in the case of Boyle, of intersubjective trust.[60] And once again, one finds that *demonstrative reasoning* existed prior to Galileo's experiments in a discourse of sexuality. Thus the scientists of the seventeenth century found themselves reinventing such ancient concepts as stoic apathy. Steven Shapin, Peter Dear, and Robert Markley argue that seventeenth-century experimental sciences invented a specific "literary technology" known as "virtual witnessing," which was adopted from such nonscientific milieus as witchcraft trials where "virtual witnessing" was utilized to produce droves of "reliable eyewitnesses." The practice involved "the repetition of experiments, the warrant of prior 'authorities,' and an overarching invocation of piety and nobility."⁶¹ Although the authority of virtual witnessing in witchcraft trials usually invoked ideological conceptions of authority extraneous to experimental observation, it should be stressed that, to the extent that "reliable eyewitness" presupposes apathy a "thousand times more exquisite" than pleasure, it is also a "pleasure" technique.

A brief look at Renaissance sexuality seems to confirm that the discourse on sexuality widely employed "collective experience" and intersubjective "trust" in the formation of "normal sexuality." One may claim that perhaps inspecting a hymen was the first rigorous method of objective verification, combining intersubjectivity, passion, morality, and perversion in a single procedure, long before Galileo used it against Aristotelian science. Or, policing sodomy before the time of Galileo required intersubjective verification, as in Venice, where authorities set up a commission on sodomy in every district of this lustful city to verify and to report to the respecting authorities all evidence of "broken asses."⁶²

The following case should further clarify this point. The patriarchal court of Venice in the 1470s heard a case by a certain priest of San Stefano who was called upon to witness—that is to "prove"—the "normal sexuality" of his parishioner, Nicolo. Nicolo's wife had accused him of impotence, claiming she was still a virgin, and therefore deprived by her husband of being a good Christian and mother. To save face, Nicolo was forced to disprove his wife. He called upon his friend Francesco to help him, who in turn called the local priest of San Stefano to be the witness. Nicolo, the priest, and a couple of parishioners went to a certain house to visit two female prostitutes. The following is the graphic testimony of the priest:

> Shortly thereafter there arrived Nicolo, named in this case along with Giacomo, scribe of Francesco Barbetta, Francesco himself and a certain old man who Francesco said was his father. Before these people the said Nicolo pulled Magdalena to him and began to kiss her. When he saw that this witness was watching him he called to him, "Come here, Priest." Then he took the hand of this witness saying, "look here, I am a man, even though some say I cannot get it up." And he made this witness feel his member, which was erect just like the member of any other man. This done, Nicolo placed Magdalena on a nearby bench and began to carnally know her. Seeing that he was being watched he called to this witness saying, "Sir, come here, put your hand here." When this witness had done that, Nicolo extracted his member and covered his hand with sperm. Seeing this the witness went over to Giacomo the scriber saying, "Giacomo, give me your hand." Giacomo took the hand sullied with sperm in his and cried, "Priest, you have tricked me." After dinner, moreover, Nicolo went to bed with the above mentioned Maria. When

he saw that he was watched he called again to this witness saying, "I hope that you will be able to testify valiantly," and made it so that he touched his erect member while he was in the act of carnally knowing the said Maria.[63]

The verificational procedures involved in this case are trustworthy friends, just as in Galileo's pendular experiments. In the case of Galileo's pendulum, isochrony is proved by collectively counting of the swings; here, passing semen among three people proves evidence of sexual "normality." Although a true gentleman does not hold the convention in high regard, the trust in evidence has been established. "Objectivity," declares Theodore Porter, "arouses the passion as few other words can;"[64] perhaps this is in part because trust in numbers presupposes the history of sexuality as much as that of truth.

Following the above argument, Neoplatonic identity is deeply implicated not only in epistemological and moral arguments but, prior to those, in arguments about asceticism and "pure love." A small part of this subject has been explored in an already-mentioned work by Mario Biagioli on the relation between new science and the sexual politics of Neoplatonic academies. Biagioli places Galileo at the center of the institution of Platonic love, the *Accademia dei Lincei*. The young Roman aristocrat Frederico Cesi established the academy on the principles of Platonic chaste love between the male members. Each member, including Galileo, Biagioli explains, must swear an oath of chastity, and was required "to avoid 'the attraction of Venus,' 'bad women and profane love,' 'venereal lust,' 'prostitutes,' 'tempting lust,' 'low passions of the boy,' 'carnal drives,' 'libidinous excitements,' and 'the body's innate desires.'"[65] Cesi did not intend to eradicate *Eros,* but to separate it from procreation and heterosexuality, to redirect it by means of homosocial bonding into a rational force so that it might be shared among the male scientists who must coproduce knowledge. Desire was thus detached from a "natural" tie with a female body and reattached "unnaturally"—that is, perversely—to knowledge, thus making knowledge the sole object of desire. But as Biagioli observes, this perverse pleasure may have as a latent function masculine empowerment. Male homosociality—bonding among men "freed" from women—proliferates male dominance over nature and tethers masculine power to knowledge, making knowledge the driving force behind the new science. Perhaps this Neoplatonic fusion of pleasure and power governed Galileo's decision to leave his mistress while in the Medici's court. As David Noble observes,

Like Augustine in the fourth century the father of modern science abandoned his mistress in pursuit of a higher calling, in a world of men. The year he left Padua, he published his celebrated Sidereal Messenger, which was an "instant success," especially when his discoveries were publicly confirmed by the Jesuits. The following year, flushed with his new fame, Galileo visited Rome, where Frederico Cesi invited him to become a member of the *Accademia dei Lincei*. Galileo's timing could not have been more perfect; a year earlier, he would probably have had to decline the invitation, handicapped as he was by his "feminine bond."[66]

Like Biagioli, Noble noted the "perfect timing" of Galileo's discoveries and professional opportunities. Conceivably, Galileo's Neoplatonic economy of pleasure and chastity has something to do with making this "perfect timing." One may argue that insofar as the rules external to the world of the senses govern both mathematics and chastity, the chaste body shares the same economy of pleasure with mathematics and thus occupies the same strategic position of dominance over the world.[67] With chastity, figuratively speaking, Galileo "mathematized" his pleasures and elevated his courtly status. It is almost as if he turned himself into a Cartesian "dead eye," which re-orders the world according to the rules of its own making.

The point of the above analysis is to stress that the Galilean scientific revolution cannot be reduced to only a conceptual revolution, unless these new concepts are related to a specific homoerotic ideology of pleasure. Any attempt to avoid this relation may lead to its reappearance in the very argument for excluding it. Koyré seems to be the case in point here. He recasts the Neoplatonic concept of the nonobjective world as *"l'Absole indetermine,"*[68] the starting point of Galileo's science, in part to inaugurate what Foucault calls a "discontinuity" in the history of science.[69] Instead of explaining Galileo's science as a moment in a longer narrative of scientific progress, Koyré explains it in Foucault's phrase, by "starting from nothing."[70] This is partly a response to Ernst Mach, for whom Galileo's discoveries and concepts induced from his observational experience help to explain science as a progressing continuum initiated by "classical empiricism." In particular Mach's claim that "Galileo did not supply us with a *theory* of the falling bodies, but investigated and established, wholly without performed opinions, the *actual facts* of falling,"[71] irritated Koyré. For Koyré, in his history of Galileo's science, Mach relied too much on naturalized history, as he did on bodily experience in his empirical science. But Koyré does the same thing, except that he unknowingly takes a Sadean turn. He reconstructs the history of Galileo's science by rupturing the "body" from naturalized history, and he looks at it as though he was looking through the Cartesian "dead eye"—that is to say, not as it happened, but as it should be reconstructed. Since Koyré's history no longer streams in a stable continuum, his critical perspective opens the body of history to endless ruptures, and more importantly, to the supramatistic pleasure of historians like himself, to an analytical *jouissance* of "nothingness" as a starting point for any radical reconstruction of history. In the light of the Cartesian revolution—for which, according to Koyré, Galilean mathematical physics is responsible—Koyré's Galileo reads as a pure object of Koyré's ascetic pleasure in the politics of shame.

SADE AND THE GALILIZATION OF PLEASURE

In *The Order of Things,* Foucault argues that an intersection between *mathēsis* and *taxonomy* unified representational schemes of "classical episteme" and so subjugated heterogeneous and diversified fields of representation to a single monolithic field. This explains how mechanics and pornography, for example, share similar schemes of representation and how this similarity mutually constitutes both fields by merging them. Following this logic, then, Galileo's natural philosophy and Sade's philosophy of the bedroom fall under the same representational schemes of *mathesis universalis* and their application unto instrumentation for the purpose of "pleasure of analysis. "While historians usually treat Galileo's pendulum only as an extension of his mathematical speculation, rarely do they see the desire and "sex" in it regulating Galileo's mathematical imagination. I suggest that Sade's use of *mathesis universalis* and *instrumentation* in his philosophy of the bedroom reveals the concealed and yet constitutive "sex" in Galileo's pendulum.

Consider the following quotation from Sade's *Juliette* which may cast a different light on the sexual fantasy of instrumentation:

> "The appointments you see here," said our host, "are alive; they move when the signal is given." Minski snaps his finger and the table in the corner of the room scuttles into the middle of it; five chairs dispose themselves around the table, two chandeliers descend from the ceiling and hover above the table.
>
> "There is nothing mysterious about it," says the giant, having us examine the composition of the furniture from closer on. "You notice that this table, these chandeliers, those chairs are each made up of a group of girls cunningly arranged; my meal will be served upon the backs of these creatures."[72]

Sade's table, chair, or chandelier, exhibits both a structure of pleasure and of mechanics. Minski snaps his fingers and the thing follows his desire. This description suggests an important parallel between the discourse on sexuality and the mechanics concealed in Boyle's literary technology of "virtual witnessing." Unlike the natural order, mechanical order is based on the principle of a human desire to construct order. In this case, Sade portrays the principle of heterosexual desire in which objects, like women, follow man's mechanical fantasies. In the functioning of the machine, Sade insists, sexuality has been reinvented, now as a pure discourse. Marcel Henaff unpacks the relation between sexuality and mechanics in Sade's use of the French word *le meuble,* furniture. In French, Henaff elaborates, *le meuble,* as a noun, means "movable piece of furniture," but as an adjective it describes something loose, soft, and easily broken, something easily constructed and reconstructed, just like desire. As Sade has shown through the invention of the "libertine body"—and here I follow Henaff's lead—every human body as *le meuble,* a passive instrument, can, by virtue of mathematical schematization of pleasure, easily become *la machine* (the active, productive, sexually pleasurable body). It can, in other words, easily become a system of newly coded pleasures. It is this schematization of pleasure by means of abstract discourse that Foucault calls "sex." When the body, as *le meuble,* has been quartered in de Sade's debaucheries to its elementary units of pleasure, then mathematically rearranged, pleasure has been amplified and codified into mechanical systems. The rules of mechanical constructions—measuring, assessing, drawing accounts of, arithmetical combinations—quantify pleasure but also sexualize objects and the discourse on "objectivity." "Sex" and mechanics have one thing in common: the "useful function" of a body upon which a discourse on sexuality may assume various forms, varying from different disciplines, such as from mechanics to psychoanalysis.

Carolyn Merchant and Evelyn Fox-Keller agree that the Neoplatonic sexualization of scientific rationality in the seventeenth century and the formation of a new mechanical science helped to structure an empirical method "that would produce a new form of knowledge and a new ideology of objectivity."[73] Bold sexual imagery, language of sexual praise such as "hard facts," "penetrating mind," or "hard-core sciences," all of which have shaped the language of modern experimental method, have also sexualized rationality by objectifying nature and the body. Sade is a prime example of the sexualization of objective rationality. By placing "objectivity" in his literary narrative as if in a scientific laboratory, Sade reveals, as we will see, two important aspects of "objectivity": that "objectivity" is a perverse pleasure, and that "objectivity" goes beyond science.

Certainly aesthetics and the moral philosophy of Hume, Smith, and Shaftesbury, their concerns for "the contemplative joy," "catholic and universal beauty," and the "virtue of disinterestedness," have contributed to the history of "objectivity." But this "ethical" history remains incomplete without acknowledging also how the pain of pleasure empowers a rational discourse of "objectivity" and turns the body of the scientist into an aptitude, a capacity for the increase and proliferation of alternative pleasures of rationality to "the weak orgasm of modestly endowed males."[74] Sade presents mathematical narrative as a lustful alternative to sex. One of the functions of mathematics in Sade's text is to transform confession into "scandalous" literature—about consummated acts, bodily touching, improper gazes, obscene remarks—and into the pleasure of the objective gaze. Male genitals are described in inches and as instruments whose precise measurement establishes the predictability of pleasure. The sexual act can be observed, "objectively" investigated, and measured. In *The 120 Days of Sodom* Sade recounts: "he sucked my saliva for fifteen minutes . . . ; turned it thousand times about in his mouth . . . at the end of three or four minutes I distinctly saw him swallow it"; "twelve inches of tongue through my anus . . ."; [75] "he was himself fucked by ten men . . . he withstands as many as twenty-four without himself discharging." In another description from *Philosophy in the Bedroom,* libertine Dolmance is sodomized by his valet in the presence of young Eugenie, who has been invited to discover these novelties: "Dolmance: . . . Ah, by Christ! what a bludgeon! never have I received one of such amplitude . . . Eugenie, how many inches remain outside?—Eugenie-Scarcely two. Dolmance: Then I have eleven in my ass! . . . What ecstasy!" Or: "Dolmance turns Madame de Mistival upon her stomach, takes up the needle and "(begins to sew her asshole). . . . Here, Mamma dear, take this. . . . And again that! . . . (He drives his needle into at least twenty places.)"[76] The pleasures of singular accounts accumulated in a summary of the balance sheet that follows.[77] As Henaff writes of this balance sheet: "Because the body is cut up, mechanized, counted, erotic relations are reduced to mere combinations and must be realized in the organization of a system of variations designated to establish the greatest possible number of connections between available bodies and their organs."[78]

These precise quantitative descriptions of the sexual act by a disinterested observer transforms sexual pleasure into the pleasure of a voyeurist glance.[79] For Sade's narrations to offer the pleasure of voyeurism he has to merge sex and *mathēsis* in his text; his narration, as Foucault points out, "must be decorated with the most numerous and searching details; the precise way and extent to which we may judge how the passion you describe relates to human manners and man's character is determined by your willingness to disguise no circumstances; and what is more, the least circumstance is apt to have immense influence upon the procuring of that kind of sensory irritation we expect from your stories."[80] Henaff relates Sade's attempts to objectivize pleasure to a new social discipline,[81] and Foucault posits in *The Order of Things* that the "universal calculus, employed by Galileo, became the general matrix of order in the "classical" age. By virtue of mathematical measurement, the pleasure of the body was replaced by the pleasure of precision, counts, and quantities.

But are Sade and Galileo really related, as the above paragraph seems to suggest? At first glance they belong to two different histories—one of pleasure and one of reason—but upon closer inspection their histories are deeply interimplicated. Galileo's mathematization of physics set the stage for Cartesian dualism and for the separation between the body and *cogito*. The body in Cartesian dualism is just another object whose reality is established only as a physical mechanism constructed according to the principles of mechanics. Mechanization of the body, as Sade demonstrates, transforms amorous relations by annexing pleasure to the rules of mathematical discourse, and establishes its reality mathematically, as it does the reality of time. Sexual fantasies and desires are enmeshed with the rules of objective representation. Unless there are numerous and exact counts these sexual fantasies could not be satisfied. A pleasure is no longer a singular experience but the aim of a sexual "project" where "pleasure arises from the mechanical mastery of bodies and a strict accounting of operations." The rational structure of mechanically arranged sexual orgies, for example, "transforms," in Henaff's words, "the relations between the subjects into relations of simply quantifiable bodies and organs." The confession of sexual pleasure turns, with Sade, into the report of an orgy which, as a sexual "project," is an *experiment* in the use of pleasure which demanded precision in order to verify and establish the confessed reality but, more importantly, to intensify pleasure by means of mathematical quantification: by measurements of sexual organs, by quantities of bodies, by measurements of sexual acts, and by drawing the balance sheet of the measuring operations. Nothing excites erotic imagination, Sade insists, as much as large quantities. Indeed, quantity in itself becomes a pleasure; according to Henaff, "quantity denotes luxury and abundance, and hence political and economic power."[82]

One could claim, then, that what Galileo was to nature in terms of mathematization, mechanization, and quantification, Sade was to sexual pleasure. As Galileo demonstrated with his pendulum that time is not a "natural" experience but a discursive construct, so Sade demonstrated that pleasure is not a bodily but a discursive construct. By virtue of measurement, a "bodily pleasure" became a quantifiable and reportable variable capable of being analyzed methodically and endlessly; like any other physical force it can be broken down, calculated, predicted, applied and controlled. "In short," Henaff writes, "the form of reason thus penetrates the matter of passion, just as in physics energy enters into the calculation of a dynamics."[83]

By making quantitative measurement a pleasure, and confession an experimental report, Sade "Galileoized" sexuality, or merged the quantitative method of measurement with representation and pleasure; pornography and science coextend their logics. Both Galileo and Sade generated their pleasures from the use of mathematics in imagined experiments. Koyré argues that Galileo's use of "mental experiment," which allowed him to order nature according to his mathematical phantasms, and Sade's use of mathematics in his "mental" sexual experiments to contrive pleasure by the use of mathematics both represent instances where mathematics orders and reorders things according to the rules of pure discursive pleasure, pleasure because of pleasure, that is perversion.[84]

There is here another important conjunction between Galileo's science and Sade's pornography explainable through reference to Baudrillard's "over-signification of the real."[85] Baudrillard was assessing the semiotic drama in the context of electronic media and the seminal role of pornography and science played in the formation of what he calls, "hyper-reality," that is the beefing up of reality by the force of the self-referential sign. The beginnings of this may be traced back to Galileo and Sade's representation of nature and pleasure as *res cogitans,* as signifying and signified thing, rather then *res extensa,* unsignified things. Galileo's project of reconstructing nature according to *mathesis universalis,* that is as *res cogitans,* demanded an "over-signification," "an *orgy of realism*" by indexing and cataloging the infinite numbers of objects. Everything has to "pass into the absolute evidence of the real,"[86] an obscene thing indeed, by which the "real" in Galilean science, as Koyré argued, may be permitted to be only a sign. But that was already where the "real" meets "sex." Sade's "terror" of sex seems to carry the same burden of the "real," whereby the violence of the sign impresses itself as the "real." "Sex" can be "real" only if measured, quantified, and catalogued by the literary narrative which, captured in words, allows for the endless inspection and analysis of its reality, a voyeuristic method, Baudrillard argues, already developed by the new science. "Science," Baudrillard writes, "has already habituated us to this microscopics, this excess of the real in its microscopic detail, this voyeurism of exactitude—close-up of the invisible structures of the cell—to this notion of an inexorable truth that can no longer be measured with reference to the play of appearances, and that can only be revealed by a sophisticated technical apparatus."[87] Galileo's pendulum or a seventeenth-century dildo are "apparatuses" invented to signify in their different ways the reality of "sex" and sexuality of the "real."

PERVERSE PLEASURES AS *RESISTANCE* AND *NORMALIZATION* OF SCIENCE

Sade conceptualizes "objectivity" as a perversion of observational indifference. In placing perversion at the root of sexual pleasure, Sade anticipates Freud inasmuch as Freud sees the pain of unfulfillable desires at the heart of human sexuality. They also differ significantly, however, in that Freud conceptualized sexual pleasure as part of a libidinal structure located in the body, and Sade concocts sexual pleasures through his narrative. Here, Sade anticipates Foucault in treating pleasure as a discursive artifact. This is evident from

Leo Bersani's account of Sade. According to Bersani, Sade does not reproduce *actual* fellatio, flagellation, and coprophagia, but produces a *pacing* of the text that is more perverse than its content.[88] He uses quantitative descriptions, theoretical hypothesis, empirical tests, mathematical schemes—that is, modalities of "objectivity"—in his narrative to contrive textualized pleasures. In his effort to recreate sex textually through elaborate and highly-mechanized orgies, cruel schedules, and cold-blooded calculations, Sade employs apathy, the spirit of "objectivity," to intensify the pleasure of pain as unnatural, that is, as a textual pleasure.[89]

"Objectivity" for Sade is a personal stand toward one's pleasures. Deleuze represents this position as "the *impersonal* element" of a "violence with an Idea of pure reason."[90] Deleuze underscores this personal position to pleasure as the basis of demonstrative rationality. The "personal" experience in Sade's narrative serves only as data for generating narrative pleasure via "impersonal" analysis—endless repetitions, quantifications, and the absence of any sentiments for those who are in pain. Remaining indifferent to pain denotes the ultimate test of disinterestedness and, if attained, is the source of highest pleasure. This severe observational coldness, as Deleuze points out, aims to desexualize the body—to relieve pleasure from the "natural" body— in this case of the disinterested observer—but not without the resexualizing of these very procedures.[91]

According to Deleuze, if Sade assumes the Kantian principle of self-determination in relation to pleasure and mathematicizes pleasures, he convincingly demonstrates that objective rationality is not free from "lust." "Coldness" as a personal/objective position "is the essential feature of the structure of perversion" which, in Deleuze's view, conjures up perversion and rationality in a symmetrical codependency: "The deeper the coldness of the desexualization, the more powerful and extensive the process of perverse resexualization."[92] In Sade's narrative, the discourse on objectivity appears as a "process of perverse resexualization."[93] In naturalized sexuality, the subject is supposed to preserve the heterogeneity of the drive by resisting the law that homogenizes and objectivizes the drive. In this context, the drive is perceived as liberating, and rules are seen as artificial and oppressive. But in contrast to naturalized sexuality, Sade's texts persistently shatter subjectivity by means of the cruel objectivity of rules. Sade denies the subject his or her natural entitlement to pleasure, but only in order to grant it to the *procedures,* to a set of conventions for gathering and explicating facts. Here the drive does not come from *inside* the subject as some natural locus but from *outside* nature, from mathematical discourse. Thus Sadean pleasure, as Henaff argues, "can itself be produced only as a movement of objectivity . . . as the plan for a destruction that will be brought about in an outside that is modeled desire."[94] Objectivity, thus, is precisely this outside scheme which on the one hand assures the conventionality of objectivity but, on the other hand, with its coldness, perverts its conventionality by becoming a pleasure.

The "perversion of objectivity" has a dual meaning here. Sade represents perverse pleasures as resisting an identity normalized by the naturalized codes of pleasure. Sade's "objectivity" is not only about objectified pleasures, but also about how these new normalizing conventions of fact gathering, conventions *outside* nature, which, due to their genealogical paradox, lead through the pain of their application to a rupture of the established codes of "natural" pleasures. It is through this perversion rather than opposition to the normalizing conventions that, in Foucault's scheme, technologies of modern subjectivity could receive a "transgressive reinscription" and resist the orthodoxy of the old science. As Jonathan Dollimore observes, Foucault promotes here a transgression over the rules and norms by perversely reaffirming their conventionality.[95] Foucault writes:

> Rules are empty in themselves, violent and unfinalized; they are impersonal and can be bent to any purpose. The successes belong to those who are capable of seizing these rules, to replace those who had used them, to disguise themselves so as to pervert them, invert their meaning, and redirect them against those who had initially imposed them. . . . So as to overcome the rules through their own rules.[96]

By realizing that there is no conventionality without perversion nor resistance to outside conventions, one thinks of perversion strategically. Perversion observes and assesses the rules, and above all, as Foucault claims, acts practically in terms of seizing the rules in order to turn them against themselves. If the rule was that the body and senses determine pleasure/truth, the perverse position of seventeenth-century science resists this conception. Perversely, not the body but discourse determines both pleasure and truth.

The effects of these strategies amount to a rupturing of the normative conditions for the formation of an agency, and to forcing autonomy, that is, asserting a domain of perverse resistance. If "objectivity" for Sade meant a destruction of subjectivity and an affirmation of pure procedures, affirmation of rational reasoning over "natural" pleasures, then he follows Foucault's scheme of resistance. Sade accomplished not only a perverse resistance to "normalized" subjectivity and the normality of pleasure by disassociating pleasure from the body and associating pleasure with the text, but this destruction, in turn, as Nietzsche later asserted, clinches the fulfillment of "asceticism" as the *painful* condition of objectivity.[97]

From this perspective, one could supplement Koyré by arguing that Galileo's science and the *eros* of negation were mutually inclusive, coextensive, and analytically interchangeable. Negative reason did not suppress pleasure, much less end it. As Foucault would insist, negative reason pushes pleasure forward, breaks new ground, marks new frontiers, spirals upwards; in short, pleasure intensifies as it nears the ultimate object. The self-referential rationality of "objectivity" simultaneously increases its inscriptive negativity. And here is the other important aspect of "objectivity." The normalization of its perverse moment in the history of science becomes a self-reproducing scheme of the historian's identity and, for this reason alone, calls for its rupture.

THREE

The Jesuits' Homosocial Ties and the Experiments with Galileo's Pendulum

In 1597, in Manila's Parish of San Ignacio, a group of Jesuit educators held flagellation exercises during Lent. A group of slaves "sang a solemn *Miserere* (Psalm 50/51) at the end."[1] In 1645, the Jesuit astronomer Giovanni Battista Riccioli decided to take Galileo to task and test his law on the pendulum. With his nine faithful Jesuit brothers, he went through a different type of bodily mortification: together and without interruption they counted in twenty-four hours, about 87,000 pendular swings. Although distant events in time and space, Jesuit flagellation and the Jesuit pendular marathon both exemplify the culture of the body as a source of higher pleasures based on celibacy and self-mortification. Historian Peter Dear has emphasized the dominance of the Jesuit discourse on mathematics in seventeenth-century experimental physics. While he emphasizes Jesuit mathematics as a vanguard of the new natural philosophy, he, however, like his counterparts, makes no reference to their bodies and pleasures.[2]

With their intellectual competitors, the Renaissance Neoplatonists, Jesuits shared love for mathematics and experimentation, finding in them pleasures of a masculine rationality and the reason for homosocial bonding. Jesuits stayed away from the paganism of Neoplatonic science and subjected themselves to a higher individual and group discipline directly inferred from the Christian discourse on sexuality. This, however, did not make their identity and rationality less dependent on pleasure, in fact one may argue that the pleasure of a masculine rationality and homosocial bonding embedded into the order of the brotherhood became far more important to them. I am prepared to argue that this may have been one of the most important factors in their becoming the scientific "vanguard" of seventeenth-century science, and that rejecting a priori reasoning as relevant for the physical world and cautiously accepting

Galileo's mathematical imagination and "mental experiments," forced them, in part, to scrutinize Galileo's discoveries with their bodies and senses and, in the process, to transform their ascetic discipline into experimental practice. Epistemologically, rather than making a radical shift as Galileo had done, the Jesuits took sagacious steps towards Copernicanism and made a gradual transformation, as Dear successfully demonstrates, from Aristotelian naturalism to mathematical physics. In this respect they cannot be regarded as being as brave as Galileo in quickly accepting the consequences of Copernican revolution. The progress of scientific rationality, however, was not the only thing on their minds.

The Jesuits' primary concern rested in the affairs of the human soul and its salvation. Their eschatology pitted rationality against pleasure. One should not lose sight of the fact that the Jesuits understood their science also as a "bridge to salvation," which was parallel to salvation via the sacraments. In the "Rules for the Scholars of the Society," the Jesuit H. Nadal articulates this clearly: "Let everybody know that society has two means by which it strives for its end: the one is a certain force, spiritual and divine, which is acquired through the sacraments, prayer, and religious exercises of all virtues and which is granted by the special grace of God; the other force exists in the faculty which is ordinarily acquired through studies."[3] Both ways lead to the liberation of the soul from "carnal pleasures" and to the desexualization of the body. Because conceptualizing science and soul-saving go hand and hand with the Jesuits, their rational will necessarily will be connected to their search for higher pleasures, bodily transformation, and celibacy. No doubt, the use of mathematics helped the Jesuits to mount a "cognitive shift," but the Jesuits' preoccupation with mathematics came along with their shift to asceticism. Bodies were foundational to the new science; Jesuit mathematical discourse must not therefore be dissociated from a more general process of desexualization of the body, and resexualization of science. Only from this perspective, moreover, it is possible to measure the comparative exclusion of women from science. Because women, presumably, were embroiled in "carnal pleasures," they were outside the male homosocial circle of scientific practice. Together, the Jesuits' "fear of fornication" and desire for homosocial pleasure actually sexualized their science and contributed to what has correctly been identified as a "cognitive shift."

JESUITS AND COUNTER REFORMATION PLEASURES

Jesuit identity was shaped in large measure by the Counter-Reformation. Political and religious disturbances of the sixteenth century resulted in the increased consolidation of a previously fragmented Catholic Church which, after the Council of Trent, mounted a vigorous but contradictory offensive against Protestantism and other heresies.⁴ By simultaneously assimilating the nascent rationalism of local or popular cultures into its dogma, the Church also rigorously reinforced its old monastic discipline and sexual austerities as a way of strengthening reason.⁵ Galileo's Counter-Reformational Italy outdistanced other countries in its return to the rigor and orthodoxy of an earlier age, which invited new ascetic orders such as the Jesuits to blossom. Jesuits, more than any other religious order, reveal the contradictions of these Counter-Reformational strategies. On the one hand, the Jesuits demonstrated the modern rationality of classical sciences and introduced it into secularized secondary education. On the other, they returned to the old methods of monastic asceticism.⁵ Rivka Feldhay elaborates the unintended secularizing consequences of the Jesuits' "reformist" strategies:

> Guided by the ideal of *vita activa* the Jesuits were gradually eroding the walls which segregated the traditional intellectual elite from the world. The Jesuit educational program entailed living with secular students, preparing for secular careers, and opening up the curriculum. New contents were legitimized, new canons of knowledge recognized, new technological possibilities imagined. The sensibilities of the traditional guardians of the faith were thus provoked. Suspicious towards any modification of the power-knowledge relation arose and the boundaries between profane and sacred knowledge were newly problematized.⁶

Copernican rationality mounted a serious challenge to the authority of the Catholic Church and its Thomistic and Aristotelian dogmas. The Jesuits believed that it was better for the Church to spearhead, rather than to prosecute, this new rationality as a way of understanding and developing proper countermoves against its enemies. This was a strategic position, but in addition, the Jesuits truly believed that scientific rationality could provide and reclaim the lost universal consensus within the evermore fragmented Church, and more importantly, between the secular community and the Church. Developing a science and a secular education based upon scientific rationality became a religious passion, one of whose by-products was to set the framework for their professional identity.

In their pursuit of rationality the Jesuits ushered in various reforms. Generally speaking, their main achievement was to initiate a process of secularization within the Church, one that took numerous forms. Acting as "reformists," Jesuits sought to build rational consensus within the fragmented Church through mediation and compromise, rather than by using pre-established dogma to judge, punish, and repress disputants. In this sense, they contributed to the development of a communicative rationality that became the basis of a new institutional consensus. Although knowledge for the Jesuits was a bridge to salvation, and discourse on rationality a tool to build a link between faith and reason, the Jesuits also developed a secular education. They separated rational studies from contemplation, offered increased autonomy for the studies of reason, developed a new curriculum, insisted on pedagogical exercises, and divided the day, the week, and the year into periods of study and of rest.[7] They divided the curriculum into specialized disciplines and thereby laid the matrix for modern education, which as we now know, gradually undermined their own conservative views.[8]

Their reformist strategies helped move the Jesuits into a new power-elite within the Church, where they found themselves in a delicate position. The Jesuits carefully balanced Dominican threats to their own "Galileanism," and balanced Galileo against their own "Thomism."[9] In terms of Galileo, the Jesuits' relations were ambivalent, to say the least. They needed Galileo to advance and test their reformist views, but they also needed to control his Copernicanism, which they hoped to achieve through open dialogue.[10] As a client of Medici, forced to produce spectacle, and please thereby a marvel-driven nobility, Galileo also sought a dialogue with the Jesuits. The patronage system of Medici's court, after all, as Feldhay points out, was in many respects incompatible with Galileo's professional interests, since the Medici were not particularly interested in Galileo's scientific plans.[11] To pursue his own scientific interests he sought a "common cultural field" with individuals equipped with the same, or similar skills and knowledge. Consequently, Jesuit mathematicians and philosophers became the natural interlocutors of Galileo. Galileo, indeed, needed the Jesuits for the legitimation of his discoveries, just as the Jesuits needed Galileo for the legitimation of their new "modernist" status within the Church's educational hierarchy. Through their dialogue, two different discourses on science, reality, and experience emerged. The dialogue crystallized their differences as well as their commonalities, creating a fluid tension between Galileo's increasing calls for the rejection of Thomistic theology and the Jesuits' simultaneous preservation and modernization of it. The tension between them ultimately lead to Galileo's trial in 1633.[12]

In spite of the conflict, a Jesuit rationality based on asceticism and Galilean mathematical physics was nevertheless inextricably, if not temporally, bound to produce "classical physics." While Galileo, in order to produce depersonalized objective knowledge, concealed the importance of his relation to his body and pleasure, the Jesuits kept it up front. Much like Calvinists, although for different reasons, they combined the old monastic asceticism with modern scientific rationality as a way to create for themselves a modern Christian identity.[13] While Calvinists sought economic control, Jesuits sought control over scriptural interpretation. Consistent with the early Christian asceticism which maintained that reason stems from overpowering carnal pleasures, Jesuits empowered their reason in relation to the body by means of sexual discipline, namely celibacy, confession, or sometimes flagellation.[14] To empower reason over carnal pleasures was a prerequisite for the sound interpretation of the Scripture, but also a source of pleasure and power over the body. To the extent that the Jesuit body was an essential part of an experiment, this empowerment and its pleasure were also a prerequisite for doing science, which makes the Jesuit economy of pleasure all the more important for examining their science. Because in the Jesuit discourse of rationality, text can seduce mind if the body is not sexually disciplined, Jesuit science is a path of salvation through desexualization of the body.

JESUIT CELIBACY AND SCIENCE

Historians agree that Jesuit mathematicians of the seventeenth century, rather than Galileo, laid the foundation for experimental physics. Peter Dear's book *Discipline and Experience* provides the most up-to-date, and perhaps the most sustained support of this claim. Drawing on Foucault's discourse on rationality, Dear divides the homogenous narrative on the history of "scientific revolution" into various discursive dialects, meticulously described and analyzed in order to formulate the Jesuits' use of mathematics to denaturalize "experience" and to promote modern "experiment" as a new scheme of experience.[15] Perhaps the most radical consequence of the Jesuits' science was, for Dear, to subvert "nature" and natural "experience" to a mathematical discourse. In stressing this point, Dear, like Koyré, argues for anti-empiricist representation of the "Scientific Revolution." He posits:

> At the most fundamental level, however, Koyré was clearly right: his crucial point was that it is impossible for nature to speak for herself. Even with novel deployments of apparatus and technique to bring about hitherto unknown behaviors, no knowledge can be created unless those new human practices and new natural appearances are rendered conceptually in an appropriate way. . . . In that sense, the Scientific Revolution was indeed a matter of a cognitive shift rather than the simple acquisition of new information that demanded a new theoretical framework to accommodate it.[16]

Stressing the primacy of "concept" over "experience" in the history of science and using the Jesuits' mathematics as the case in point, Dear, like Koyré, promotes a Cartesian exclusion of "bodies and pleasures" from the history of scientific rationality. But, whereas Koyré subordinates the seventeenth-century discourse on mathematics to his own discourse on transcendental rationality,

Dear, as an ethnographer of discourse, asks how the Jesuits' understanding of their own discursive practices leads to the formation and formulation of their revolutionizing concepts.

The Jesuit mathematical discourse, Dear emphasizes, denaturalized "experience" by subordinating its naturalized appearance to mathematical rules. Starting with the Jesuits and ending with Newton, Dear argues that "experience" became a discursive construct rather than an extension of the body and senses. But the transformation of "experience" as body to "experience" as a discourse, according to Foucault, had been already initiated by a Christian discourse on "sexuality," which aimed to subjugate pleasures to the rules of rationality. With Christianity, Foucault stresses, "sex" became the discursive code of an "experience" and a property of a discourse rather than of the body. In other words, because scientists have bodies and pleasures and have them according to the rules of their histories, the "experience" of sex became the first "experience" to be constructed discursively, and it was on the tail of that discursive transformation of "experience" that scientific experience could become "discourse" too. From this vantage point, the history of Jesuit science must be read alongside the history of their bodies and pleasures. I will examine them in that order.

The Jesuits' study of science was not a process of collecting information but a "force" to achieve personal salvation, which in the sixteenth and seventeenth centuries became a motivation for experimental science. By the end of the sixteenth century, the interest in experimentation led to the formation of the first academies of experimental science, such as the *Accademia del Cimento* in Florence, the Royal Society of London, and the *Académie Montmor* in Paris.[17] Often, the Jesuits' school disciplinary organization served as a model for other academies of science. In the words of historian John Heilbron, the Jesuits were "the single most important contributor to the support of the study of experimental physics in the seventeen century."[18] The seventeenth-century Jesuit science was versatile in its scope and ambiguous in its worldview. Dear claims that the Jesuits represented the core of the "physico-mathematical vanguard of a new natural philosophy," and in the words of William B. Ashworth, they also had a particular "zest for experimental science."[19] The Jesuits, he observes, "were interested in every newly discovered phenomenon, from electrostatic attraction to the barometer to the magic lantern, and Jesuits played a major role in discovering many new effects on their own, such as diffraction and electrical repulsion." Their commitment to a disciplined way of life endowed Jesuits with "a keen sense of the value of *precision* in experimental sciences"[20] which made them, rather than Galileo, determine the swing of a second of time and accurately measure the rate of acceleration for freely falling bodies. With their superb mathematical skills, observational patience, pedagogical imperatives, and scientific instrumentation, the Jesuits were best able to perform, scrutinize, and advance Galileo's experiments, rather than the *Accademia del Cimento,* whose purpose was to "follow [Galileo's] method of research and of criticism."[21] In their constant search for the grounds of universal rationality, the Jesuits mixed their Aristotelian view, that sense-experience grounds scientific rationality in certainty, with mathematics. They viewed experiment as the perfect fusion of observation and mathematics. Instrumental demonstration, it was claimed, produced universal and public experience as mathematics did, "embodied in the structure of the presentation."[22] Father Riccioli was a proponent of this view. In his experiments, Father Riccioli demanded that experience be universally evident; otherwise the experiment can always be doubted.[23] For him, mathematical measurement provided universal evidence of experimental experience. Indeed, he was the first to actually measure the second of a swing of Galileo's pendulum.[24]

Collaboration was the Jesuits' way of doing science. The Jesuits collected and distributed most of their work through correspondence. Operating as a disciplined and world-wide missionary society, they covered a wide range of natural and artificial phenomena, observed, described, and published their results in an encyclopedia, which was disseminated and made ready for analysis.[25] In this respect, in the words of Ashworth, "the Society of Jesus, rather than the *Accademia del Cimento* or Royal Society, was the first true scientific society."[26] The Jesuits developed disciplinary knowledge before others, professional discourse on the question of scientific objectivity, and above all—by mixing mathematics with other sciences—transformed Aristotelian "common sense" experience into disciplinary discourse.[27] All historians agree that the Jesuits' science was not the work of a solitary mind, but the collective effort of dedicated men.

But there was another "cultural field" Galileo shared with the Jesuits. The Jesuits, just as Galileo and other members of *Accademia dei Lincei,* used their "bodies and pleasures" to advance their science. Biagioli points out the similarities between the academy and the Society of Jesus. The disciplinary structure of the order and its sexual politics served to a certain extent as a model for the academy. Like the Jesuits, the members of the Academy sought each other as philosophers/soldiers. By caring for each other and excluding women and heterosexuality, they, like the Jesuits, shielded each other from the temptations of the body in order to allow reason to spread its wings in *co-knowing*. The adding of the founding father of the order of St. Ignatius Loyola to the list of the patron saints of the academy attests the extent to which the Jesuit spirit dwelled therein. However, according to Biagioli, there were some important differences. Like the Jesuits, Academicians saw the mind as inextricably bound to the body and to sense experience. But whereas the Jesuits sought the mortification of the body and asceticism, the academics subjected their bodies and pleasures to a pragmatic "'drill,' an ongoing training for a very philosophical life."[28] While the

Jesuits exercised Christian asceticism, the academics seemed to exercise some form of antique *askēsis*. In any event, their sexual politics towards women reflects their shared understanding of knowledge as a homosocial desire.

As Biagioli has mentioned, this description of early scientific organizations fits quite well with Eve Kosovsky Sedgwick's claim that male homosocial bonding—whether expressed in a military barrack, a club, or a laboratory—is a method developed by men to collectively mobilize and redirect their desire away from women and towards each other in order to advance male power/knowledge. Celibacy was central to forming Jesuit homosocial bonds. In the Counter-Reformation, Jesuit celibacy manifested a personal will to salvation as the ground of universal certainty, which empowered the Church, to combat pagan naturalism and Protestant subjectivism. Historically speaking, celibacy emerged in monasticism and increasingly became important for the nascent power of the Church fathers after the fourth and, more significantly, after the sixth century.[29] By the time that the Society of Jesus was formed (1545), the celibate priest was the rule rather than the exception in the Catholic Church. The Church historian Charles A. Frazee concludes, "From the seventeenth to the twentieth century, the law of clerical celibacy was accepted and enforced to a degree never before witnessed in the history of Christianity."[30] As celibacy became a law of the Church so misogyny and sodomy became its counterpart. Not only was celibacy related to the imposition of procreative sexuality as the only legitimate mode of sexuality, but it also—at least according to some historians—constituted an alternative mode of pleasure.

John Boswell's pioneering work on the history of Christianity and its adoption of the Roman military homoerotics, places the Jesuit Order in this tradition. In his view, the Jesuits inherited homosocial bonding and homoerotics from the militaristic ethos of early Christianity. Although Christ warned that those who live by the sword will die by it, he ordered his angels into armies, and he himself said that he will bring not peace but the sword to earth. This Christian military ethos, inherited from the hierarchical and disciplinary structure of the Roman military, culminated with the Crusades, Templars, and Hospitallers, and the masculine companionship of the military became the model for the "Christian soldier," who in this mutually supportive company, battles the demons of his inner soul. Bodily discipline and sexual austerity "purified" the male soul but also intensified the pleasures and power of male bonding. Envisioning Christ as the head of a great army of male archangels, angels, and saints, the clergy perceived male companionship as a divine relationship emanating directly from Christ who held the entire heavenly army together through homoerotic desire. Thus it was not unusual to find same-sex pairing among the military saints. Christian art represented this relationship as the fruit of pure love. Artists, for example, invented two Theodores—the General and the Footsoldier—always depicted as a pair, often with their arms around each other.[31]

Young Augustine was not a stranger to this feeling. He captures the pleasures of male bonding when he praises it over the "pleasure of the bed." For him, male companionship opened new and more satisfying channels of desire. He comments:

> To talk and laugh. To do each other kindnesses. To read pleasant books together; to pass from lightest joking to talk of deepest things, and back again. To differ without rancor, as a man might differ with himself. . . . These, and such like things, proceeding from our hearts as we gave affection and received it back, and shown by face, by voice, by eyes, and by a thousand pleasing ways, kindled a flame which fused our very souls together, and, of many, made us one.[32]

In Augustine's view, celibacy is not the absence of pleasure, but a male's higher pleasure, which kindles a flame in men to feel as "one." Under the ever-increasing pressure of celibacy for the clerical elite, the newly coded economy of pleasure gradually emerged. On the one hand, the religious elite ensured for itself homoerotic discourse as its economy of pleasure; on the other hand, procreative sexuality became increasingly imposed upon women, and homosexuality was prosecuted by the Church as sodomy. For the Church, a discursive homoerotics empowered the Church fathers to define women in terms of "carnal pleasures" and to secure by virtue of this discourse a dominance over women.[33] Born out of a Christian military ethos and its imperfect homoerotic discourse, the Society of Jesus inherited homosocial structures of power and homoerotics as a semiconcealed discursive pleasure, from which their scientific inquiries never broke away, though they were transformed into a discourse on scientific rationality.

Ignatius Loyola, the founder of the order and a former military general himself, exemplified in his instructions for spiritual exercises not only the importance of homosocial bonding for the life of a Jesuit, but also how homoerotic desire becomes a discourse on rationality.[34] With his homosocial prerequisites, he envisioned the Society of Jesus as a military organization, led by a spiritual general whose purpose was to facilitate the new identity of a Jesuit as a Spiritual soldier.[35] Noah Porter describes the homosocial structure of this Society:

> Ignatius had been a soldier, and he carried all the soldiers into his new order. He aimed to bring the ardor, the daring, and above all, the discipline of the camp, to do their utmost in the service of the Church. The name of the head of the order was General. All the gradations and divisions were

military. The authority of each superior over his subordinates was complete and despotic. Every member, from the highest to the lowest, vowed the most implicit obedience to any and to every order from the General. It was obeyed on the instant, whether it reached them by day or by night, in sickness or in health.[36]

This homosocial identity was based on celibacy, reason, and a battle against the bodily pleasures.[37] Unlike the classic minds of the Enlightenment—Galileo, Descartes, Newton, and Kant—who regarded reason and rationality as a universal, bodiless principle of truth, Loyola regarded Jesuit reason as the outcome of the Jesuit's struggle with his pleasures—among other things, his struggle with sexual discipline. For him, reason was the soul's weapon to win its autonomy in the battle against the irrational forces of bodily pleasures. In the struggle over one's body, reason emerges in one's mind, one's feelings and one's will, and directs the conscience from sin to truth. Rational truth is not only a depersonalized statement of facts, but above all a personal achievement in relation to one's sins. Loyola believed that a soul, that which is a subject of pleasure, finds in reason an ally against the pleasures of the body and also the source of a new bodiless pleasure. Disciplinary operations such as confession, self-surveillance, and penitence disassociated pleasure from the body and associated it with reason.[38] This new economy of abstract pleasure aimed at the formation of a new identity. In this sense, Jesuit rationality stands for the struggle for new subjectivity. As Loyola wrote in his *Spiritual Exercises:*

> In the persons who go from mortal sin to mortal sin, the enemy is commonly used to propose to them apparent pleasures, making them imagine sensual delights and pleasures in order to hold them more and make them grow in their vices and sins. In these persons the good spirit uses the opposite method, pricking them and biting their consciences through the process of reason.[39]

Clearly, Loyola treats rationality and reason not as a Cartesian disembodied *cogito,* independent from *eros,* but as a technique of *eros*.

Homosocial bonding and its implicit homoerotic desires presupposed for the Jesuits a struggle over the body. Its disciplinary effects were a source of scientific rationality. For Loyola, reason is represented as masculine asceticism in opposition to feminine weakness to pleasure. In the twelfth rule of his spiritual exercises he casts *asceticism* as an exclusively masculine rationality, suited only to the manly soul, or the soul which found in reason an ally. "The enemy," he writes, "acts like a woman, in being weak against vigor and strong of will."[40] Such a "genderization and sexualization of God" is, in the words of Elizabeth Schussler Fiorenza, "biased" and "sexist-exclusive"[41] and reveals sometimes conscious, sometimes unconscious gender-exclusive practices common to the domination of one gender over the other. Clearly, the Jesuits' science belongs to the history of Western science in which, according to David Noble, gender exclusive language has evolved for women in a particularly grim direction.[42] But for Judith Butler, the exclusion of women from homosocial discourse on rationality is more than a biased position.[43] The absence of women from the production of power/knowledge, affirms their absence of any position in a phallocentric culture. For Lucy Irigaray, this is not a deviation from the system but the very essence of its productive power based on sexual differences. She writes:

> Sexual difference would constitute the horizon of worlds more fecund than any known to date—at least in the West—and without reducing fecundity to the reproduction of bodies and flesh. For loving partners this would be fecundity of birth and regeneration, but also the production of a new age of thought, art, poetry, and language: the creation of new *poetics*.[44]

Production of female identity along the line of sexual differences inevitably proliferates woman's subordinate position as her identity and as her pleasures. Thus, because of sexual differentiation she had no place within the discourse on scientific rationality. This differential scheme would have allowed the Jesuits to deduce from it, as if given a priori, woman's identity in terms of "carnal pleasures."

Furthermore, this scheme incorporates but also complicates the Jesuits' invention of their identity as modern scientists by putting "women" in between them as that which they mutually exclude in order to be "one" with each other. This homosocial dynamic between Jesuit men is primary for defining women's identity only in terms of what men have to denounce in order to bond, that is bodily pleasure. The Jesuits excluded women not simply as "Other," but also as an art of reproducing a symbolic system of power centered around male bonding. As Butler points out, the subject and the excluded "Other" are masculine byproducts aiming to close the phallocentric economy and totally exclude the feminine.[45] The impossibility of being an "Other" gender, and being only an absence of identity, emphasizes the self-referentiality and self-fulfillment of the schemes of sexual differentiation throughout Jesuit science. The Jesuits' homosociality also exhibits the self-referentiality of this scheme. By defining reason in opposition to an invented "feminine," the self-referentiality of sexual difference not only strengthens male homosocial ties, but more importantly, by associating pleasure with the *masculine,* eroticizes reason, increasing both "aptitude" and "capacity" for a "cognitive shift" in science, that is, a shift to a strictly homosocial economy of pleasures.

THE JESUIT BODY AND GALILEO'S PENDULUM

Galileo's Copernicanism mounted a serious challenge to the Jesuits' science. His law of free fall, claiming that the time of the free-falling body was independent of weight and distance, undermined the Jesuit's admiration for direct observation in physics. Galileo's pendulum was the case in point. In a simple and publicly available way Galileo's pendulum allowed anyone to witness the clash of paradigms. Although Galileo formulated the law of free fall, he never went through the *pain* of providing a precise measurement of the time of the falling body. This omission by Galileo is due partly to the absence of the precise measurement of time and partly because free-falling bodies descend so quickly that it was impossible to directly observe regularities, which required an extremely high point on the objects' release. All trust was placed in mathematics. However, there were other possibilities for achieving a precise measurement of time, and for testing Galileo's claims. One was theoretical, which the Dutch mathematician Huygens advanced. By inventing the theory of the curve along the pendulum's swing, he determined what part of the curve had isochronic motion, allowing him to invent a pendular clock. The other way to test Galileo's law demanded strong will and bodily discipline, more than theory. To test the isochrony of Galileo's pendulum empirically and see whether or not it could reliably measure free falling bodies, long, tedious, and exhausting observations were required. When Father Riccioli decided to take Galileo's pendulum to the task and test its laws, he had at his disposal both disciplinary resources and his order's passion for precise time.

"Riccioli's historical accounts," Dear points out, "describe an exercise in calibration as an illustration of procedure, rather than representing discrete experiments the findings of which were to contribute to an understanding of nature."[46] Father Riccioli, in other words, did not learn through these experiments anything new about nature, but he did perform calibrational *exercises*. In contrast to Galileo's "mental experiments," the Jesuits bring their disciplined bodies into the mathematical schemes and perform experiments. Their will to knowledge, in other words, does not exclude their bodies. Koyré is quick to stress this experiment as very important for the development of classical physics by putting an emphasis on "precise measurement," rather than on the pre-epistemic disciplinary conditions of the Jesuit will. Certainly mathematics played an important role in advanced seventeenth-century experimental science, but Riccioli's experiments clarify the view that extraordinary will and discipline are essential too.[47]

Even though failing to theorize, Koyré gives a fascinating, although according to Dear, not always reliable account of the discipline involved in Riccioli's experiment.[48] Only sleepless nights, passion for precision, self-determination, and the synchrony of chanting turned the Jesuits' bodies into the first experimental clock. To achieve the precise measurement, it was necessary to measure the swings of a pendulum in comparison with the motions of the planets. This required a long period of observation which, in turn, involved particular skills with the occasional pushing of the pendulum to keep its oscillations constant and consistent in its path.[49]

In his experimental design, Father Riccioli selected a period equivalent to six hours on the sundial and counted short pendular swings for these periods. His results did not conform to Galileo's predictions, so he decided to repeat the experiment with the help of other Fathers whose names Koyré records: "Stephanus Ghisonus, Camillus Rodengus, Jacobus Maria Palavacinus, Franciscus Maria Grimaldus, Vicentius Maria Grimaldus, Franciscus Zenus, Paulus Casarus, Franciscus Adurnus, Octavius Rubens." Letting Koyré's dramatic narrative take over from here, the second trial went as follows:

> Moreover, Riccioli recognizes that for his aim the sundial itself lacks the wanted precision. Another pendulum is prepared and "with the aid of nine Jesuit fathers," he starts from noon to noon: the result is 87,998 oscillations whereas the solar day contains only 86,640 seconds. . . .[50]

Koyré observes that Father Riccioli repeated the experiment one more time:

> Disappointed yet still unbeaten, Riccioli decides to make a fourth trial, with a fourth pendulum, somewhat shorter this time, of 3 feet, and 2.67 inches only. But he cannot impose upon his nine companions the dreary and wearisome task of counting the swings. Father Zeno and Father F. M. Grimaldi alone remain faithful to him to the end. Three times, three nights, 19 and 28 May and 2 June 1645, they count the vibrations from the passage through the meridian line of the Spica (of Virgo) to that of Arcturus.[51]

. . .

Now it is obviously impossible to use so rapid a pendulum simply by counting its swings; one has to find out some means of summing them up. In other words, one has to construct a clock. Actually it was a clock, the first pendulum clock, that Riccioli built for his experiments. Yet it would be difficult to consider him a great clockmaker, a forerunner of Huygens and Hook. His clock, indeed, had neither weight nor spring, nor even hands or dial. As a matter of fact, it was not a mechanical clock, but a human one that he built.

> In order to sum up the beats of his pendulum Riccioli imagined a very simple, and a very elegant device: he trained two of his collaborators and friends, "gifted not only for physics but also for music, to count un, de, tre . . . (In the Bolognese dialect in which these words are shorter than in Italian) in a perfectly regular and uniform way, as are wont to do those who direct the execution of musical pieces, in such a way that to two pronunciations of each figure corresponded an oscillation of the pendulum." It is with this "clock" that he performed his observations and experiments.[52]

Koyré concludes:

> Yet even if we admit—as we must—that the good fathers corrected somewhat the actual result of their measurements, we have nevertheless to acknowledge that these results are of a quite surprising precision. Compared to the rough approximation of Galileo himself and even to those of Mersenne, they constitute a decisive progress. They are certainly the best ones that could be obtained by direct observation and measurement and one cannot but admire the patience, the conscience, the energy, and the passion for truth of the R. P. Zeno, Grimaldi, and Riccioli (as well as of their collaborators) who, without any other instrument for measuring time than the human clock into which they transformed themselves, were able to determine the value of acceleration, or, more exactly, the length of the space traversed by a heavy body in the first second of its free fall throughout the air, as being equal to 15 (Roman) feet.[53]

The work was so rigorous and tedious that several fathers dropped out; only two fathers faithful to Father Riccioli continued. Due to one's man devotion to another, common to members of an "essentially military" society accustomed to "the discipline of the camp,"[54] the experiment could proceed. Indeed, counting tens of thousands of oscillations is a "dreary and wearisome task," which requires extraordinary discipline and concentration on the monotonous mechanical detail of a pendular swing over a period of days. It requires, in other words, the patience of a monk which, to quote Foucault, makes "possible the meticulous control of the operations of the body, which assured the constant subjection of its forces and imposed upon them a relation of docility-utility" which "had long been in existence—in monasteries . . ."[55] and where precision is the "fundamental virtue"[56] of a discipline.[57]

To establish and standardize the second of time was another case of the Jesuits' proclivity for precision.[58] Their interest in precise time was political and religious as much as scientific. The punctual observance of matins in monasteries, as we have seen, was among the first incentives for living by the clock. Although Koyré stresses the epistemological significance of "precise time" in classical physics, he conceals the disciplinary history of time. Koyré does not register that grounding the concept of time in their bodies, their discipline, and their ascetic space, along with their exceptional knowledge of mathematics, altogether amounted to the birth of "classical physics."[59]

The Jesuit discipline of bodily synchrony and temporal precision common to their *mantic art* was particularly suited for Riccioli's experiment. Experiment is not a pure artifact of mathematical discourse, but also, in the case of the Jesuits, a lucky convergence between the mechanics of a scientific instrument and the exigencies of the Jesuitical discipline. One finds in Loyola's *Exercises* an insistence on this temporal ordering of activities essential for Riccioli's experimental design. There are structural and individual reasons for this. Structurally, disciplinary methodology—regulating everything from daily schedules, postures, and diets to ceremonial procedures—involved the organization of time. A total temporal mapping of activities achieves two aims: it *seals* the Order from any intrusion of events from the outside world, and also allows the synchronization of everybody's activities up to a second. Loyola advises that synchronized activities for the near future should be already planned in the present; when going to bed, for example, one should prepare for the morning, in the morning one should prepare for the noon, etc. This total temporal mapping, Roland Barthes observes, evokes a "machine-like" phenomenon in the fullest cybernetic sense of this term.[60] To synchronize rhythmical activities was the essence of the Order's discipline. To map these activities allowed the Order to be active in the outside world by remaining hermeneutically sealed from it. On the individual level, Loyola's *Exercises* subject identity to the rhythmical form of a language. Loyola recommends "praying in rhythm by joining a word of the *Pater Noster* to each breath."[61] This is a method, according to Barthes, familiar in the Middle Ages under the name of *Lectio divina,* "which consists in breaking the name down into its etymological albeit whimsical components (*Di-os,* he who has given us life, fortune, our children, etc.)."[62] Loyola, unlike Buddhists who decompose names to eradicate all nominal connections, Barthes posits, "recommends an exploration of all signifiers of a single noun in order to arrive at a whole; he wants to wrest from the form the whole gamut of its meanings and thereby extenuate the subject—this subject, which in our terminology is endowed with a pleasing ambiguity, since it is simultaneously *quaestio* and *ego,* object and agent of the discourse."[63] By rhythmical and repetitive chanting, the prayer is, Barthes argues, both denaturalized and reconstructed. To use language in mantic art is not uncommon to religions, but whereas other religions use language as a docile conduit of spirit, Loyola's method of prayer is different. It aims to endow it with the status of a "second code," itself a source of ecstasy.

Once again, Barthes elaborates, "it is a matter of the technical elaboration of an interlocution, i.e., a new language that can circulate between the Divinity and the exercitant."[64] Rhythmically spoken language synchronizes the individual with a discovered code. It acts as a matrix to order one's soul, to discipline the body, and ultimately, to achieve divine reason.

The ecstasy of the Jesuit language induced by the rhythms of the "second code" rather than by the body, demonstrates what Julia Kristeva, in her celebrated book *Desire in Language,* calls "the rule of the Phallus." Following Lacan's theory of the symbolic, Kristeva argues that a symbolic order of language that denies a connection with the maternal body and replaces it with the phallic symbol simultaneously repudiates femininity.[65] "The 'subject' who emerges as a result of this internalized repression," Butler summarizes Kristeva's point, "is necessarily dissociated from his own body as well, a subject whose unity is purchased at the expense of his own drives, and whose denial is renamed as *desire*."[66] Following Kristeva's argument then, subjugating the speech to the "second code" brings the body to the stage of pure desire, here a creative force for bonding among the men. Semiotic aspects of language such as the rhythms, pacing, breath or polyvalences of speech reveal, we learn from Kristeva, the workings of the prediscursive body as a "handmaid" to the symbolic order and to "the rule of the Phallus." Taking this route from a code to the body, Kristeva hoped to rupture the phallocentric order of signification and open the speech to poetic and polyvalent meanings as a way to de-eroticize the symbolic and eroticize the body suppressed by language. If Kristeva's semiotics represents a rupture of the phallocentric language and the emancipation of the feminine subjectivity, then the Jesuits with their subjugation of the speech to the "second code" represent a reinvention of homosociality.

Let us dwell for a moment on the latter. The Jesuit practices of rupturing subjectivity and their repetitive practice of experimental counting, merged here in their exercises. Repetition, according to the mantic art employed by the Jesuits, ruptures the importance of a natural subject and is therefore transformative of identity. After the four weeks of the *Exercises,* Loyola explains, transformation through de-formation of the Jesuit's identity occurs. The practitioner must be submerged in repetition in order to rupture the old and give birth to a new identity. "He," Barthes writes, "is to repeat what depresses, consoles, traumatizes, enraptures him in each narrative; he is to live the anecdote by identifying himself with Christ: 'To demand sorrow with the sorrowing Christ, laceration with the lacerated Christ.'"[67] To count "un, de, tre, . . ." for 2 x 87,998 swings as the Jesuit experimenter did, is a mode of self-annihilation. According to Barthes, however, such laceration basically implies pleasure, a "jouissance."[68] By using this word, I mean to imply that the repetitive chant must have been at once an erotic, spiritual, physical, and conceptual transformation.[69]

The experimental synchrony of Riccioli's experiment reveals an array of bodily knowledge external to mathematical discourse and yet important for experimental science. Counting the speed of a pendulum is a disciplinary *operation,* not a formula. The speed of a swing demands focused eyes, ears, and informed minds. In short, it demands a disciplined body. The collaborative focus on synchronic swings produces a synchrony in seeing and hearing, the counting procedures. Thus for the Jesuits, mathematics applied to experiment was not of purely epistemological origin. In mathematical physics, as we will see in the second part of the book, experiment demanded a subjection of one's body to the abstract rules of mathematical discourse, while the pendulum's experimental task to homogenize this personal experience could only have been achieved among bodies historically developed out of particular homosocial discourses and homoerotic desires. While Galileo's call for local hands, eyes, and minds to collaborate an experiment, Father Riccioli and his monks verify the accuracy of Galileo's pendular measurements by stressing a will to knowledge and, no less important, bodily discipline.

The Jesuit's body, then, like his rationality, was shaped by a discourse on sexuality as much as by a discourse on experience, and in this respect the body's economy of pleasure must be recognized in the history of their science. In the case of Riccioli's experiment, of course, commitment to a brother was more than just a friendship: it was a passionate devotion to another man, part of their vocational and "professional" conduct, and an individual pursuit of the pleasure of masculine discipline. In addition to this, the Jesuits, more than other groups, used their fraternal networks to distribute their works and findings at a time when scientific channels were not in place. Once again, Father Riccioli's experiment demonstrated that testing Galileo's transcendental claims could not have happened without simultaneously testing the boundaries of our own bodies.

The Jesuits demonstrated that the pleasure of rational conduct and eroticized reason by means of misogyny were the formative components of experiment as a collaborative enterprise. The solidarity of Jesuit experimenters also points to the underlying economy of pleasure in camaraderie, as both the social structure of an experiment and as its erotic resource. The passage through the meridian line of the Virgo to that of Arcturus, in Father Riccioli's experiment, was indeed a cosmic event of planetary motion thought, dissected into the smallest component and summed up by means of the social grace of synchronized chanting among the three monks. The "human clock," in other words, was built upon the social grace of these ascetics, of their collective chanting into "a perfectly regular and uniformed" piece of music.[70] This relationship of camaraderie, while necessary for the experiment, also had an unreflected and pleasurable outcome in transforming "homosocial" ties common to monks and soldiers into structures of objective knowledge.[71]

Finally, any history of science that emphasizes the mathematical foundation of modern experiment conceals the fact that experiment meant subjection to the conditions of experimental design. For Riccioli and his monks to test Galileo's assertions also meant *testing themselves bodily* in a new rational discipline of scientific experiment. Their experiment was thus also a means of self-formation through self-discipline, which reveals a very different approach to scientific method than the one given by Galileo and later by Descartes. The Jesuit "method" suggests endurance, patience, prudence, silence, wisdom, and temperance, all of which are needed for experimental work. In addition to the theoretical a priori, we should also consider that a body, pleasures and discipline are a practical a priori of an experimental science. In their view this has been missing from Galileo's pretentious claims that, "Without making the experiment I am *sure* (italics-mine) that the result will be as I am telling you . . . ," and that mathematical phantasm as a pure discursive pleasure can substitute bodily practices in experimental physics.[72] To get an actual result was to establish an experiment for which one must have a body, a will, and the ethos of an ascetic pleasure. Experimental practices could not be instituted unless and until experimental techniques themselves, the result of professional skills, were acquired through bodily exercise. Because the Jesuits channeled their "will to knowledge" in part through the body as a practical, rather than only through mathematics as a discursive, a priori, they were already prepared for the rigors of experimentation. Similar to the way in which mathematics challenge ordinary perception, so did Riccioli's experiments challenge the ordinary will and the body of a scientist. "Methodically controlled and supervised" conduct, asceticism and modern experiment, are coextensive in the history of Jesuits' experimentation. The Jesuits' homosocial effort to calculate the pendular length of a "second"—which the ancient Sumerians who, perhaps as much as the Jesuits, cherished measuring time, named *gesh* and, which incidentally in addition to a time unite it also meant "man" and "erect phallus"—demonstrates the erotic conditions of modern experiment and science.[73]

In the first part of this book, I have provided historical fragments that allow some historical generalization about seventeenth-century science and sexuality. Not writing a coherent, or even "true" story on this subject, bending connections between the fragments and the elements within the fragments allows me to deploy "repressive de-sublimation" as a discursive strategy to force a recognition of a history of science as an *unrecognized* production of the "real" by means of erotic improvisation conjured up by the analytical deprivation of sexuality. "Sex," exposed as a sign engaged in the act of writing its own history, leaves the body-work, the pendulum, and the body link disengaged from the terror of canonic pleasures. Fully aware of the historic significance of the body-instrument link and of "sex" as its representational sign, we will now examine this link by means of a pedagogy, that is, by knowledge eroticized by the body and related practices freed from "sex."

PART TWO

Pedagogy

FOUR

The "Body-Instrument Link" and the Prism: A Case Study

In order to problematize the history of Galileo's science and represent it as unrecognized erotic improvisation, so far I have emphasized Nietzsche's claim that every knowledge rests on an injustice done to the body in the name of "timeless ideas," and Foucault's call for a method that would vindicate the body.[1] But much too often we fall under the spell of this erotic improvisation by initiating another cycle of abstractions with the hope of "retrieving" the body.[2] To a certain extent the first part of this book falls under this spell. There, where I develop historical models, I represent the "body" as yet another in a discourse, albeit as a sign strategically positioned to resist normalizing canons in the history of Galileo's science. But this is only half of my task. The other half is to introduce a *pedagogy* of the body-pendulum link as bodily associative move. By taking pedagogy to mean what it meant to the Greeks—a way of transmitting a knowledge from body to body that is the body eroticized by knowledge—rather than what it means today, a theory of teaching. I now turn to the body as pedagogy, not as a natural locus but as a *practical* link.

My prime example of *pedagogy* as hitherto defined, *knowing with an eroticized body,* is a study done by Harold Garfinkel on an intersexed person with a male anatomy passing as a female, known at the Department of Psychiatry at the University of California at Los Angeles in the late fifties of the twentieth century under the pseudonym "Agnes." "Platonic realists" of professional expertise who define sexual identity through scientific discourse, are mostly puzzled by "Agnes" and have focused primarily on the medical, endocrinological, and psychiatric aspects of her case. On the other hand, "Agnes" astonished Garfinkel with her accomplished appearance as a female, thereby directing focus of his study primarily on "her" acquisition of the practical knowledge of "passing." Garfinkel launched his inquiry into her case with his tape recorder, documenting her life history, her own assessments of her identity, her desire to become a women and, most importantly, her knowledge about "passing." Moreover, Garfinkel did not regard this as an ordinary ethnographic study but rather a pedagogic moment in Agnes's body-work. In the conclusion of his study Garfinkel claims that because Agnes's production of a gender identity ensues from her sustained study of the organizing features of ordinary settings, and because of her possessed practical knowledge, she should be regarded as "the practical methodologist." By this Garfinkel means that "everyday events, their relationships, and their causal texture were in no way matters of theoretic interest for Agnes." To consider her world as a discursive construct common to professional knowledge rather than as an upshot of practical and circumstantial skills, would have been for Agnes like learning how to swim in a classroom instead of water, that is "just words" rather than practical action. Garfinkel asks: "What does a 'practical methodologist' contribute to the understanding of social order?" "What is the status of the practical knowledge *vis-à-vis* professional knowledge?" In opposition to the professional authorities, Garfinkel proclaims "Agnes's" methodological practices should serve as "our source of authority for finding, and recommending study policy" on normally sexed persons. However, this authority and these policies contradict normative expectations of gender-related knowledge jealously guarded by professional status and professional vanity, both of which are neatly braided into the single response of a prominent psychiatrist to Garfinkel's study:"I don't see why one needs to pay that much interest to such cases. She is after all a very rare occurrence. These persons are after all freaks of nature." But acknowledging in the psychiatrist's remark a Cartesian division and domination of professional over lay knowledges, Garfinkel responded by saying that nobody could have solicited "a more common-sense formula," dramatically reversing the meaning of "sex expert" in favor of "Agnes's pedagogy."

How can the status of a "practical methodologist" be obtained in relation to Galileo's pendulum? How and which body is to be eroticized by knowledge about Galileo's pendulum? And above all, how can one dismantle the fence of Foucault's discourse and move on into pedagogy? Answering these questions would require a discussion of the rules and their relations to discourse and the body-instrument link as conceptualized respectively by Foucault and ethnomethodology.

FOUCAULT'S "BODY-INSTRUMENT" LINK AND THE QUESTION OF RULES AND DISCIPLINE

The concept of a "body-instrument link"[3] first articulated in *Discipline and Punish,* illustrates, for Foucault, the channel of a microphysics of power and the meeting point between a discourse on rules and the body. In his early work on prisons, Foucault represents the body-instrument link as normalizing negative power. In his late work, however, specifically in his analysis of the techniques and instrumentations used by "queer" communities, he represents the "queer" body involved in the practices of pleasure as governed by rules somehow extrinsic to theoretical discourse.[4] There he concedes that perhaps his early discourse on rules and on the body need to be reformed.

Foucault discusses the "body-instrument link" in the context of the origin of modern subjectivity out of the body's relation to instruments of power. In the body of a soldier or a monk, both defined by subordination to a code of conduct, Foucault finds the vanguard of modern disciplinary power. Not only did the rules of conduct define monks' and soldiers' bodies, but their "souls" as well. A dialectic tension between subject and object, internal and external, ceased to exist in the body-instrument link. While in premodern societies, Foucault speculates, the body-instrument links were primarily intuited, in the time of the Enlightenment they became monitored and regimented. Here is where Foucault introduces the discourse on rules. Foucault claims that through the specification of the rules of conduct, which allow the body to be a more effective and precise object of control, anonymous power identical to these rules regiments the body and its identity.[5]

To illustrate how rules form the "soul" by disciplining the body, Foucault singles out military drills as a case in point. Military drilling requires the internalization of a code, such as bringing "the weapon forward in three stages," so as to create a collective and purely synchronized spectacle "meticulous[ly] meshing" instruments and the bodies. Foucault refers to a military ordinance from 1766:

> Raise the rifle with the right hand, bringing it close to the body so as to hold it perpendicular with the right knee, the end of the barrel at eye level, grasping it by striking it with the right hand, the arm held close to the body at waist height. At the second stage, bring the riffle in front of you with the left hand, the barrel in the middle between the two eyes, vertical, the right hand grasping it at the small of the butt, the arm outstretched, the trigger guard resting on the first finger, the left hand at the height of the notch, the thumb lying along the barrel against the molding. At the third stage, let go of the rifle with the right hand, which falls along the thigh, raising the rifle with the right hand, the lock outward and opposite the chest, the right arm half flexed, the elbow close to the body, the thumb lying against the lock, resting against the first screw, the hammer resting on the first finger, the barrel perpendicular.[6]

Military drill, Foucault maintains, links the body to the instrument in prescribed, controlled, and more efficient ways, effectively de-naturalizing the body *via* an "instrumental coding of the body." This code breaks down the total gesture of the body into two parallel series, in canonically fixed order, which correlate parts of the body with the parts of the object. "Over the whole surface of contact between the body and the object it handles," Foucault writes, "power is introduced, fastening them to one another." This link, he stresses, constitutes "a body-weapon, body-tool, body-machine complex," a new form of disciplinary power which is far more controlling and far more insidious than those forms of power which demanded only signs, products, or labor. "Thus disciplinary power," Foucault concludes, is a "coercive link with the apparatus of production."[7] This link, presumably, is simultaneously a *productive* and *transformative* subjection; it is through the military drill and the automatism of instructed actions that European societies of the "classical age" "'got rid of the peasant' and gave him 'the air of a soldier,'" or imbued "grace" and an aesthetics of mechanized movement into his awkward body.

Foucault's early fascination with the aesthetics of negative power led him to represent discipline as a simple effect of the code of power. While here the principle of aestheticization of the body belongs to institutional codes, in his later work he realized that principles of the aesthetic of power belong to the acting subject. In comparison to his late work, these early texts suffer from what Harold Garfinkel might call being "invariant to all exegesis" of *how* the code sinks down into the muscle of a soldier. In other words, they lack a reference to marching as a structured strain and to the corresponding practical pedagogy without which there could be no self-reflexive governance. Because Foucault assumes the code as the necessary element for the soldiers to be organized into a march, or perhaps from a polemical desire to denaturalize the existential and phenomenological body, he takes the code to be the historical a priori for explaining bodily discipline.

One might object that in order for the code to become a "coercive" link, the code must still be *intuited* by the individual body, which is not to suggest that the body must be a "natural" locus external to any discursive scheme. The point is simply that in order for the rules of marching to be intuited, different and informal sets of schemes, not rendered by the discourse of code, must be in play. How, in other words, does the soldier discern the "texture of relevance" in every new move through which he must maintain the coherency of the code? As a former soldier myself, I am aware that not only do all soldiers not march the same, but there are soldiers who cannot march at all, even though the parameters of military power are very clear. Moreover, those who can march demonstrate vastly different marching "styles" within the same code of conduct, and some are simply better than others, all of which suggests that while every marching solder has access to the code, the code has no access to any body. I myself remember the bodily struggle of marching while I was in the military. The rules, in other words, did not simply merge with my muscles, arms, eyes, and ears; the code was decidedly not inscribed as a tattoo over my body. Obviously, the soldier and Foucault understand the code of marching in very different ways, the former with the body and the latter with words. This kind of "body knowledge" escapes Foucault's antiphenomenological narrative; or to put it another way, he subordinates these schemes of materialization to a concept of discipline as "object without context."[8] Where Foucault sees a priori, the soldier sees his body, skills, and intuition. This is not, I believe, a trivial difference for analysis. Foucault, one could argue, collapses two different uses of a rule into one, or, to be more precise, he presents the discursive rationality of the code of marching as if the marching of a single body is its pure effect, or to paraphrase Judith Butler, as though it were a "citation" of the code. For social scientists such as Alfred Schutz these two forms of rationality, discursive and bodily, are epistemologically irreconcilable, because they operate upon two different sets of rules.[9] The first follows the rules of textual coherence; the second, the rules of contextual coherence.[10]

Foucault's bold move to define the body by virtue of the body's subjection to the code of conduct, was, doubtless, an important challenge to both existentialism and phenomenology, but while successfully denaturalizing the body, Foucault perhaps never conceptualized the rules that governed the disciplined body as the body's referent independent of his discourse. Yet these rules, too, from an ethnomethodological perspective, are "subjugated knowledges" not easily confined to a "code"—in this case to the "code" of his own discourse. Take, for example, Ludwig Wittgenstein's attempt to teach mathematics to a peasant girl in the Austrian Alps. She failed to follow his instructions; he smacked her; still she failed to follow his instructions. Wittgenstein's attempt to impose these rules upon his student seems *external* to understanding the rules of mathematical reasoning; no amount of brute force allowed his pupil to follow the rules in mathematics as though they were a code to be imposed, but patient and endless *exercises* would. Without denying that the "body" is a shorthand term for "rule-governing," one must carefully consider Wittgenstein's claim that there are *no* rules for how to follow a rule. One simply has to understand the rules by *acting*. Certainly, Foucault did not conceptualize power *only* as a brute force, but he also did not conceptualize rules other than as a power. At least in this early work, Foucault subjugates to his history of discipline the *pedagogy* of such rule-governed practices, without which no rule can ever be followed.

Foucault, however, changed his understanding of rules as codes after he began to consider "queer" practices and the instrumentation of pleasures. Because his own subject-position has changed here, he allowed the critique to work upon himself. "For a rule of conduct," he now claims, "is one thing the conduct may be measured by, this rule is another." Rules in discourse differ from the rules in conduct. This important distinction undermines Foucault's earlier discourse on rules as codes: "Given a code of actions," he now writes, "and with regard to a specific type of actions . . . there are different ways to 'conduct oneself.' . . ." Imposition is no longer the agency which regulates disciplinary practices but rather, discipline works through "a long effort of learning, memorizing, and assimilation of a systematic ensemble of precepts, and through a regular checking of conduct aimed at measuring the exactness with which one is applying these rules."[11] Disciplinary practices come out of a *pedagogy* of self-reflection through which one, by acting and observing oneself, learns ways of conduct to produce oneself as an ethical subject, and whose modifications allow for self-stylization and the "arts of existence." At the heart of this self-reflective ethics is not a coercion but a *pedagogy* that freely regulates life by means of learning rules of conduct that were always there but unrecognized beneath representational schemes. In this new context of a pedagogy of practices, Foucault formulates a new meaning for formal rules of conduct. Referring to the Greeks' prescriptions for sexual austerity, Foucault concludes, "These texts thus served as functional devices that would enable individuals to question their own conduct, to watch over and give shape to it, and to shape themselves as ethical subjects; in short, their function was 'ethno-poetic,' to transpose a word found in Plutarch."[12] One wonders if Foucault's method in *The History of Sexuality* functions as a historian's "ethno-poetic."

THE PEDAGOGY OF THE "LINK"

A well-informed philosopher, after reading the preceding section, written in the form of instructions for practical action rather than in a series of discursive analyses, has objected that this section is "hard to follow even for a philosopher." If a philosopher, that is, the most capable among all intellectuals to comprehend complex meaning, fails to understand something, then the text must be unintelligible. In other words, understanding by means of active body rather than just reading becomes an obstacle even for a philosopher, a master of a discourse.

The philosopher's complaint seems to suggest "understanding" not to be a mental process but rather the body habituating to a text. When called upon to engage in ways of knowing other than reading, the philosopher realizes his or her practical inadequacy. To know Galileo's pendulum with the body, one has to get involved with an instrument. This involvement demands a change of the subject position and its relation to knowledge. Because understanding comes through practice the analyst must relay on *a posteriori* evidence rather than on a priori claims. Sitting, reading, understanding through the lull of the temporal order of a flowing narrative, the reader's body stands as a mausoleum of the disciplinary history of the body. The *knowing-*

body, on the other hand, breaks the pleasure of the narrated time and interjects the boredom of the local time of its action. Perhaps this is the real challenge for the philosopher whose pleasures are in the text. But there, where the philosopher stops, where the lucidity of words ceases and the power of subjugation of the discourse of rationality diminishes, the *knowing*-body comes out of the rupture of a discourse grounded in a habit of the body being subjugated to the text.

The concept of the knowing body may remain an abstraction as long as we do not answer the question about rules, their locations and their elucidation. While Foucault's claims of ethics, power, or knowledge, can be settled discursively, the question "Which rules govern the body-instrument link?" cannot. To answer this one must first establish an actual link, find, and use an instrument. To represent the body-instrument link not in a historical context but as a practical pedagogy requires a change of the genre of writing too. The text is instructive, dry, technical, in a word ascetic. But the ascetic nature of this pedagogical and ahistorical text performs an important historical function. It removes pleasures from reading and returns them to the body-instrument link, that is to pleasure of pedagogy, of practices.

The point is to analyze the "body-instrument link" not as a discursive sign but as an *instructed action*. To do this I would use my own body as a source of knowledge and a specific instrument (a prism) in order first to establish a specific link and second, by means of ethnomethodology, to transmit the found knowledge about the "body-instrument link" to the body of the reader.

I invite the reader to use the prism, a simple and instructive instrument, in order to put oneself into the practical perspective and make his or her link a reflexive link with the prism available for analytical reflection.[13] Assuming that the reader has obtained the prism, the first step is to put the body and the prism in a ready-to-use mode. This can be done by simply familiarizing oneself with the instrument, holding it, rotating it, looking through it, and so forth (see figure 1). If the reader does not have a prism then he or she is deprived of the equipment with which to read this text as a series of instructions about how to discover the body-instrument link, not as a generic concept but, as his or her *use* of the prism. *Having* the prism is the first step towards establishing and knowing the link with it. By *holding* the prism the reader should realize that it already *sets* the prism in a mode of being ready-to-use in order to accomplish a certain assignment, for example, looking through it. As the reader places the eye on it to look at or through it, he or she simultaneously defines the prism as *an object with an assigned use* (see figure 2). While having the prism ready-to-use, the reader should ask what it takes to make it an optical instrument. But his or her answer should be instructive rather than definitional. Let's say the reader has

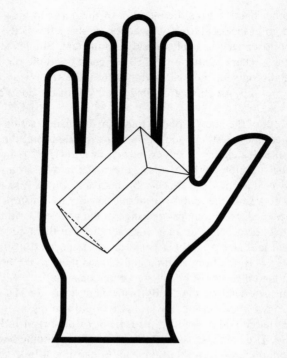

Figure 1

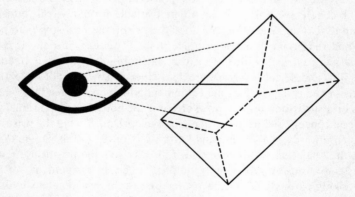

Figure 2

Figure 3

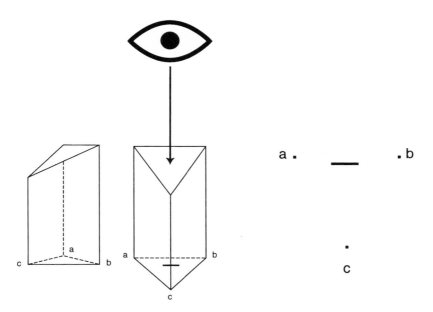

Figure 4

to find this reflective pattern (see figure 3) by looking through the prism at the line marked below (see figure 4). The prism's three inside walls work as mirrors in which the line is symmetrically reflected into a variety of patterns. How to find this pattern is a matter of finding a certain position of the prism and a certain angle of peering through it. Consider figures 1–4 as an instruction on how to achieve this practical certainty.

Looking through the prism while rotating its three reflecting walls around the marked line and seeing its reflections in a geometrically patterned order makes it an optical instrument. Individual angles enable seeing some and limit seeing other reflecting patterns. For example, only under certain angles is the reader able to see the pattern shown in figure 5. Figure 6 will demonstrate this by describing the required position of the prism and the angle of peering through it as a practical prerequisite for finding it. The prism has to be positioned vertically so that the slanted side can be used as a "window" into the reflective field of the prism (see figure 7). This is the first part of the requirement; the second part is that the marked line must be parallel with one of the prism's walls, as shown in figure 8, in order to see the pattern through the "window." But it can also be that some readers hold and look through the prism in a way that prohibits them from finding it. If such is the case, one cannot assume that something is wrong with the prism but, rather, with its assumed use (How to hold the prism, how to rotate it, where to look, etc.). The inference is that the reader encounters *the use of the prism as a reflexively produced contingency for the opening of the optical phenomena (field in physics)*.

The prism organizes the reader's work as well as the physicist's hand, eye, and neck in a way that allows the body and the prism to "stream" into one another. The prism becomes incorporated into the habitual movements of the body, becomes an extension of the body, and the body becomes instrumentalized, an extension of the instrument. This becomes a natural way of finding a sense of inquiry. At the same time, this "streaming" becomes the practical foundation for the apodictic evidence of the discovered phenomenon. The "streaming" incorporates the reader's peering orientation into the practical requirements of the prism and, as an achievement of it, the prismatic reflective patterns are found as objective knowledge. The mastery of these practical contingencies I call *praxioms,* or *practical axioms*.[14] Praxiom stands for only those actions through which the phenomenon is seen. In that sense praxioms are practical foundations in any instrumental work in "Galilean physics." Our task is to elucidate them.

Having the prism as an object ready-to-use enables the reader to "dwell" in it as an instrumental extension of his or her embodied perception.[15] By "dwelling" in it, the eye is prismatized and the prism becomes an extended body, thus setting the prism into a searching mode. The reader's

Figure 5

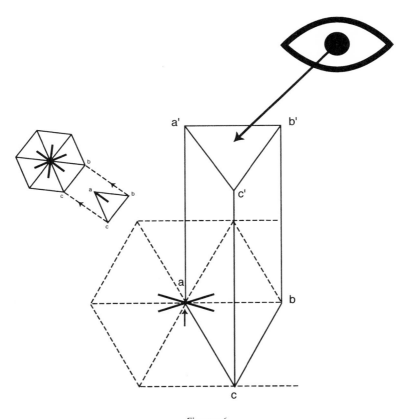

Figure 6

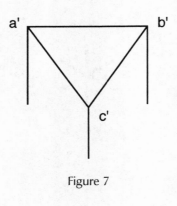

Figure 7

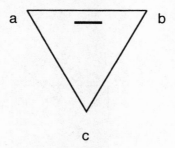

Figure 8

embodied perception is not a personal but praxeological achievement, though one that is always individually realized. In the course of finding the practical sense of the inquiry "how to rotate the prism?," "where to look?," etc., the reader is guided by his or her own sense of symmetry. The symmetry is a practical project at hand; it is in the eye of the practitioner and is perceived naively not as a project guiding the inquiry, but as the objective property of a reflective field. Once the project is accomplished, the remembered rules of such accomplishment become a path of how to find it again and again as it was for the "first" time. In other words, the order of reflection is rendered through the body/eye being instrumentalized into the proper angle of seeing the project as the objective property of the field. *The experimenter's body ought to be attuned to the practical requirements of the instrument to reveal the physical phenomenon.*

Having the instrument ready-to-use, assuming the use of the prism as the problem at hand, and instrumentalizing one's body, are some of the aspects of the practitioner's orientation that reveals the prediscursive intelligibility of the body-instrument link. For the reason hitherto mentioned the perspective is inseparable from the practical requirements of the instrument's use and the purposes of producing objective knowledge. The intuitive orientation to the instrument itself, is itself a practical object in need of reflexive inquiry. The following section in this text is dedicated to the systematic analysis of these practical objects and their so-called "transcendental rules" as effects of the reader's own achievements.

THE TRANSCENDENTAL RULES OF GEOMETRICAL OPTICS AND THE SITUATED ACTION

The triangular structure of the prism's inner walls or mirrors creates reflections of images according to the rules of symmetry. A line, or a dot, when looked at through the reflecting walls of the prism, has been multiplied and symmetrically distributed by three inner mirrors into multisided geometric figures. A line or a dot, or any other marked sign, is the given object. Its reflections and geometric objects created out of these reflections are not, however, given, but objects *achieved* through the reader's eye-prism link. Although the inner logic of these figures corresponds to the rules of geometry as *codes* of reflective formation of prismatic images, they are not separable from the prism-at-hand and the situated action of *looking* through it. Thus, only in theory are codes given a priori; in practice they are achieved and come as a posteriori evidence of a specific body-instrument relation. When Foucault imposes a priori a code of marching as rules for the situated action of marching, he mystifies both by separating them discursively. For the body and in a situated action, this separation does not exist.

One can demonstrate this point by returning to the prism. It has been assumed that the reader has used the prism in order to find designated patterns. If the reader was unsuccessful in finding them, he or she will be instructed how to find them in the following section. In the case that the reader was successful, then he or she will be able to compare his or her practical order of finding them with the one in the text. The purpose of this segment is to demonstrate that the order of prismatic reflection has a definite practical logic to which a geometrical description of the reflective pattern is ornamental and not essential. Describing this practical definiteness is also explicating what, in the course of the lived inquiry, becomes the intersubjective logic of just "one more time through." In a practical sense, this demonstration requires mastering practical contingencies. This of course presumes that the universality of the inquiry and the praxiomatics of the instrument's use are strictly confined to the particular prism described here and to an ordinary body. Other prisms and bodies would have other practical requirements.

Nevertheless, this instrument with its universal practical requirements opens the elucidational field of the instructively respecifiable logic of practical findings, its intersubjective structure as a locally lived achievement ready for the reader's reflective inquiry.

With the following exercise the reader will be introduced further to the phenomenal field of the body-prism link as the practical object ready-to-use for analysis.

Instructions: Find the following patterns (on the left side of the figure 9) positioning the prism over the line (given below on the right side of the figure 9), and peering at it through the prism. It is anticipated that the "naive" reader will have great difficulty following these instructions until reading further. However, it will be instructive to try to do so at this point in the text.

The following instructions describe how to find a pattern by peering at the line through the prism and by rotating it.

Instructions:

1) To find pattern *a,* put the prism's base (ABC) on the line so that the line cuts one of the base angles (60° each) into two 30° angles (see figure 6).

    Note the difficulty of finding simultaneously the phenomenal field and the "correct" angle of view. It is not as obvious a viewpoint in practice as it is in the text. The instructions and illustrations can indicate a discovery, but only with the prism does the sense of the instructed action and its "proper" object become intelligible.

2) To find pattern *b,* reposition the prism's base so that the line will be parallel to one of the base's sides (see figure 10).

    Note that this instruction presupposes an accomplished sense of "finding" ("Where?" "Oh, here!") as well as the competent use of the formally explicated prism ("base"; "one of the base's sides"; "parallel"). Once this is mastered, the instructions become surprisingly easy, but only after it is mastered.

3) To find pattern *c,* rotate the prism from the previous position clockwise to the point where the line will cut one of the base's sides by approximately 30° (see figure 11).

    Note the drawing for pattern *c* is a formalization, not a concrete description. The reader may find that he or she cannot see all at first. It may be necessary to peer around and turn the prism. One has to "excuse" the interrupted character of the thing at any moment of its apprehension.

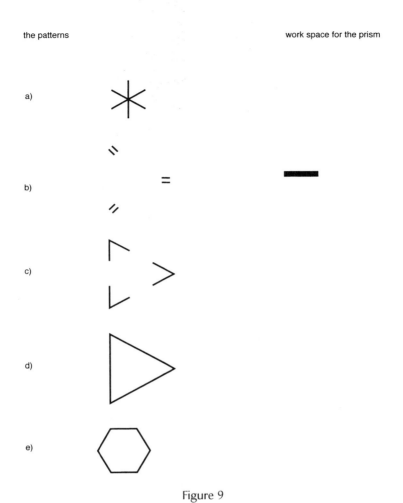

Figure 9

4) To find pattern *d,* rotate the prism's base from the previous position further clockwise while putting the line at 90° on one of the base's sides and 30° on the other (see figure 12).

    Note that the inadequacy of the instructions is betrayed by the "mirror" shifting, focusing, restarting . . . all of which make up their real-time realization. An observation can be made: the adequacy of the instrument is revealed by the "mirror" adjustments, and so forth, that enable their real-time realization.

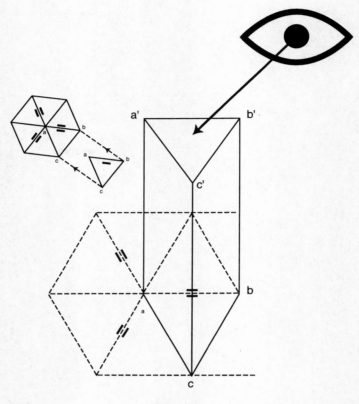

Figure 10

5) To find pattern *e,* rotate the prism's base from the previous position further clockwise so that the line cuts two of the base's sides at an approximately 60° angle (see figure 13).

Note how the achieved witnessing of pattern *d* is instructed by the prior achievement of patterns *a–d*. Further play enables the discovery that the order patterns *a–e* is revisable, not canonical. We can reconstruct the entire procedure in its sequential order by using the series of hexagons and marking the corresponding position of the prism's base below them (see figure 14).

After, and only after, the sense of the formal instruction has been discovered through the production of the appropriate prismatic gestures, does each of the patterns become an ironic description of an accomplished operation. The pat-

The "Body-Instrument Link" and the Prism 97

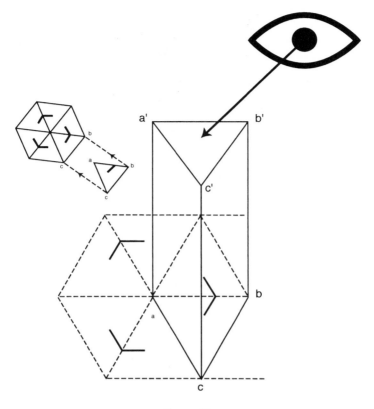

Figure 11

tern emerges in the course of the practice as the *aim, criterion foil,* and *product* of the project at hand. Finding the patterns has a definite temporal structure: each sequence builds off prior sequences, including the wrong sequences and right sequences, and these produce a spatio-analytic structure—with the star and the hexagon marking different phases. This connection provides instructions on how to find a particular pattern: If you begin at pattern d or at e, in the first case, you rotate clockwise down, and in the second case, you rotate counter-clockwise up. However, this is not the only way to find these patterns. There are other routes available. For example, the reader can also derive an e pattern from a slight modification of b. This is how to get it:

Position the prism's base as shown in figure 15 and then move the prism upwards while maintaining the side of the prism's base (AB) parallel with the given line.

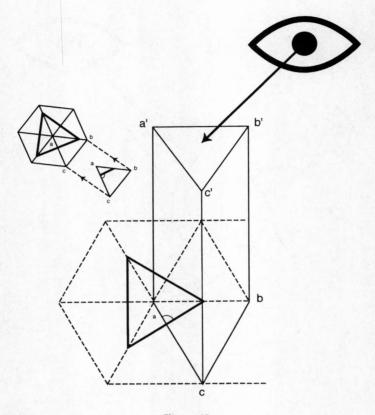

Figure 12

The following section instructs the reader on how discover a sense of mathematical objectivity of symmetry and pattern as the instrumentally achieved order of the prism's competent use. The series of instructions can *become* "literal descriptions" of the practical operations for finding the particular reflective patterns in the prismatic field. A prismatic pattern is produced through the reader's intuitively found sense of symmetry. A geometric order of the prism's reflective field thus "emerges" as an objective fact. Explicating how this sense is achieved in and through the reader's own practical standpoint is the goal of this exercise.

The instructions described here are not operations themselves. They remain to be achieved through the reader's own practical attempts. Instructions should be understood as a form of performative text that gives only practical clues as to how to operate the prism in order to find a designated pat-

The "Body-Instrument Link" and the Prism

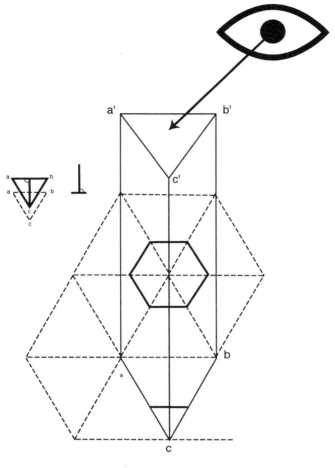

Figure 13

tern.[16] In other words, instructions require an intuitive orientation and a competent deployment of the instruments before they become actual operations of the entire phenomenon of "Galilean physics."

We called the completed actions that led the reader to actually see the designated patterns *praxioms,* or practical-axioms. Praxioms are a way of speaking of the apodictic evidence of symmetry and pattern achieved in the instrument's field. As noted earlier, I argue that these practical operations, ethnomethodologically speaking, in their achievable character, are a perspicuous case of the unexplicated foundation of objective knowledge in "Galilean

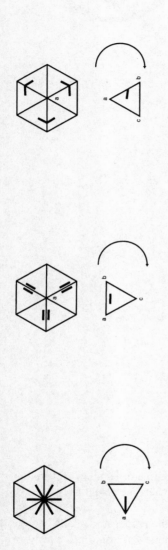
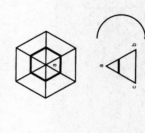

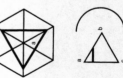

Figure 14

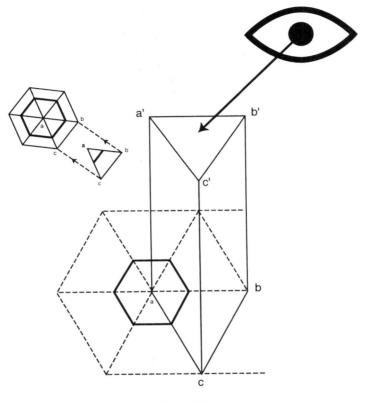

Figure 15

physics."[17] The elucidation of this foundation is inseparable from the reader's ability to master the operation that enables him or her actually to see the pattern. The mastery itself is the foundation being ethnomethodologically elucidated with the proto-scientific perspective contained in it.

FINDING THE RULE OF REFLECTION

The above instructions give a formal account of different praxioms. Praxioms, as practical axioms, are intuitively achieved, yet with a formal and respecifiable order of practical relations among the prism, the line, and the reader's way of looking through it. Mathematics in this case is a formal account of these practical necessities, and is neither the "ontology" of the prismatic field nor the projection of a transcendental ego. The reflective field of the prism is a natural phenomenon belonging to optics with all of its lawful structures. It is formulated as follows: "The angle of incidence of the parallel rays of light unto the flat surface is equal to the angle of reflection of the same parallel rays of lights." Its geometric order is symmetrical and is exhibited in the prismatic field as regular geometric patterns. On the basis of the analysis performed with the present text, we can say that the order of the reflective field of the prism is an achievable property of an instrument-organizing inquiry. In prism-organizing work one recovers the lived conditions for the physicist's idealizations, mathematizations, and formalization. The following paragraphs should demonstrate their practical foundations.

In the body-instrument link, rules and theorems "exist" in an abstract space, and yet they constantly assume the "guise" of instrument-organizing work. They exist as a formalized body of theorems associated with an inherited and maintained tradition that operates as a "logical machine."[18] In this chapter the invented rules and theorems are treated in exactly the opposite way. They are redeemed from the status of strictly formal abstractions by incorporating them into the body of a local inquiry. Accordingly, the mathematical rule in physics, and it is to be trusted to our body-prism link, is inseparable from a series of instrumental operations. My assumption is that the instrument-organizing work produces the law in experimental relationships.

In order to find the rule of reflection we can select arbitrarily a single position of the prism and analyze it. The selected position is the one described in figure 16. The prism is positioned vertically over the line on the paper so that the (AB) side of its (ABC) base is cutting the line at approximately 60°. In the triangle (see figure 16) there is a description of this position. Put the base of the prism exactly over the triangle and peer through it. An interrupted triangle pattern *c* should be discoverably apparent. The field of reflection consists of a single triangular prism's base (ABC) and its perpendicularly extended mirrored walls (see figure 17; AB–A'B'; AC–A'C'; CB–C'B'). The three mirrored walls of the prism reflect the line captured in relation to the prism and its internal mathematical arrangement. The symmetrical pattern is also extended laterally toward an "infinite" horizon, that is, it is indefinitely extended and repeats beyond the field of view, like an array of tiles (see figure 18). One has to imagine a reflection of the line's position extended beyond what is viewable by peering through the prism. Figure 19 describes the visibly reflective field of the operation described in figure 16. One notices that the line is reflected in the three mirrored surfaces, and as a result a distinct gestalt pattern becomes visible. A notation of the triangle (ABC) is also reflected symmetrically so that the initial notation of the triangle is reflectively extended in all the other reflected triangles. One has to imagine that this pattern is extended into an infinite plane that is not completely visible from the angle of view. The angle allows for only so much to be seen. However,

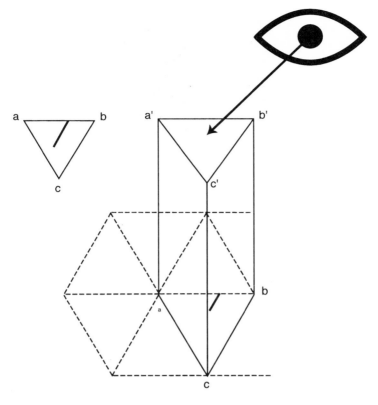

Figure 16

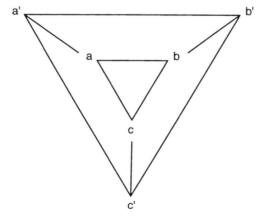

Figure 17

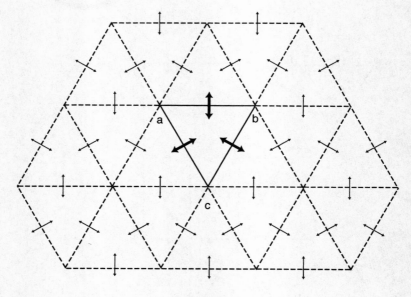

Figure 18

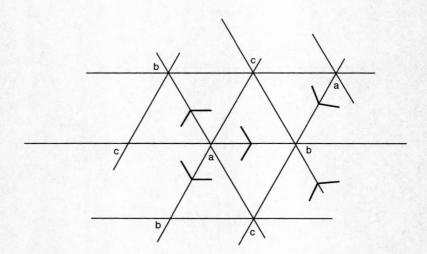

Figure 19

from what is viewed the invisible domain of the reflective field can be logically reconstructed. If we are to extend the reflective field beyond what is visibly available, the operation of reconstructing the invisible domain of the reflection will imply a rule for extending the pattern.

If the reflected base of the prism has equal angles A, B and C, then the entire reflective field is a repetition *ad infinitum* of its geometrical order. What is seen only partially, or not seen at all, is logically incorporated into the repetitive structure of the same reflective order and therefore can be reconstructed via triangular syllogism. Note that this geometrical order can provide a criterion for noticing the imperfections of the prism currently in use. The hazy degradation of iterated reflection, the distortions and smudging of the surface, and so on. The use of geometry to cast imperfections into relief implicates a *mathematical* order that differs from the visible, reflected field and its extension *as the field seen* (an expended, hazy, degrading visible field, as opposed to a "tiled" structure of identical elements extending indefinitely). For example, if we see only two angles reflected and the third extends beyond the visible line, then it can be "seen" logically by simply inferring that if angles AB are seen then the invisible angle must be C, or if angles AC are seen then the invisible angle must be B, and if we have BC then A, and so on. The same logical structure carried circularly reconstructs the invisible domain of the reflection. In this way, the phenomenon of the reflective field is given in its observable and its logically reconstructible parts (see figure 20).

After identifying the rules of the notational reflections by which the space of the reflective field is described as logically consistent with the visible order of reflection, the possibility for reconstructing the rule of the reflected position of the line follows. The question, then, is how can the law of reflection be described as an operation? The practical character of the prism's reflection is an achievement of the inquiry into the instrument-organizing work. If the prism's base is rotated and positioned over the line as in pattern *c* (see figure 16), then this particular position is the primal element for the emerging repetitive symmetrical structure of this very same position. In order to find the mathematical rule as an operation of the particular symmetrical repetition of the position of the prism, we would be intuitively guided by a "tool" in the search for a pattern. This tool can be made first by copying the prism's base (A, B, C) on a piece of cardboard paper (it needs to be thick paper for the purpose of rotating it over the page) and then cutting out the marked triangle with a pair of scissors. The second step is to mark the position of the line on the triangle and cut it out leaving an empty space, wide enough that the head of pen can pass through it for the purpose of marking the line on the paper. The angles should be marked as they are marked in the text.

To describe the tool for finding the rule of reflection: the object-organizing theorem is a triangle, given as an icon of the position explicating pattern.

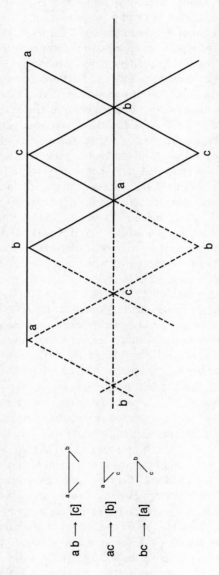

Figure 20

The "Body-Instrument Link" and the Prism   107

Each mark on one side is the reflection of the same from the other side. The cutout space in the triangle represents the line and it is left up to the reader to fill in the line, as a way of determining its axiomic position in the reflected field (see figure 21).

To discover how it is reflected, we need first to position the triangle as it is seen through the prism (see figure 16). (The reader should leave the prism in this position in order to compare it with the findings.) Mark the sides of the triangle and label them with the letters ABC from outside so that they correspond to the same angles marked from the inside. From this starting position we will flip the sides of the triangle over in order to identify the axiomic distribution of the symmetrically repetitive structures. "Flipping" is an object-organizing inquiry into the setting (tool, paper, the flipping) of the inquiry itself, and thus an "axiomic" operation, an action generating a lawful regularity (see figure 22).

If we flip triangle (ABC) with the marked position of the line on its side (A-B), then the triangle will rotate angle C 180° (see figure 23). While holding the triangle in this position, mark the line and the sides of the triangle. Peer

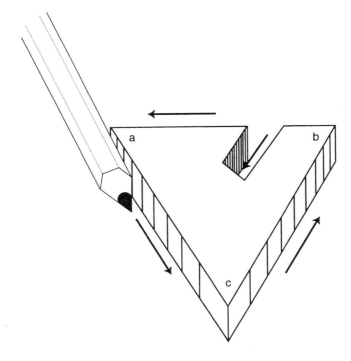

Figure 21

through the prism and compare the outcome of the rotation with the actual reflection. The new position of the triangle is an exact description of its reflection. One can carry the same operation endlessly in any direction. For example, if we flip the triangle over angle A five times then the following order will emerge from this operation as shown in figure 24. The hexagon centered A is marked. Now we can see a hexagon centered in B. Put the triangle in the starting position. We then flip the triangle over angle B five times to achieve the result shown in figure 25. The hexagon centered in B is marked. Now we can see a hexagon centered in C. Put the triangle on the starting position. We then flip the triangle over angle C five times and figure 26 emerges out of this operation. With this last operation the hexagon centered in angle C is completed. Other routes are also available. For example, if we start from the first position and flip over the A-B line, the B-C line, the A-C line, the A angle, the B angle, and finally over the C angle, each of the flips lands the marked triangle in the theoretically determined space (see figures 27a and b). The angles of the mirrored walls as the given internal mathematical relation of the prism's sides exhibit the unique property of the instrument in the pattern itself. It is not only that this prism organizes the search for the pattern but it also forms and distributes the pattern along its internal mathematical order. The latter is a static aspect. A dynamic aspect is the rotation of the prism itself. It is an intuitive inquiry into the inquireable properties of the rotating prism. The inquiry into the inquireable properties of the rotating prism has a distinct mathematically accountable operation. These two components bring a unique pattern.

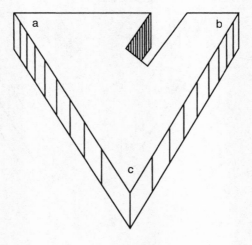

Figure 22

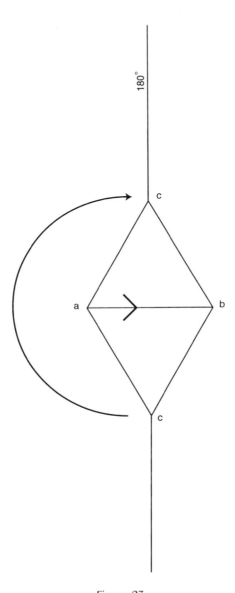

Figure 23

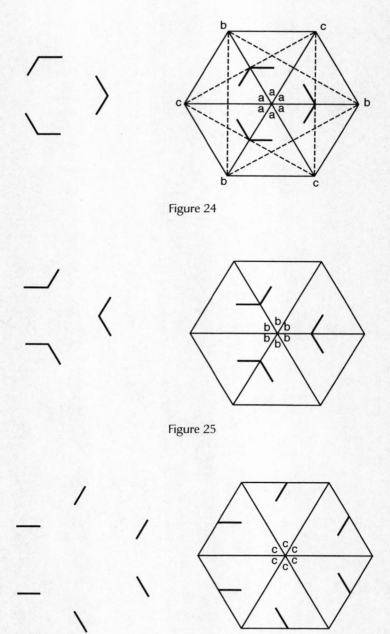

Figure 24

Figure 25

Figure 26

THE TRIANGLE IN USE AS THE THEOREM-ORGANIZING OBJECT

The triangle notated with the angles and the positioned line is the theorem-organizing object. That one side serves as the mirror image of the other already identifies the simple-observable order of reflection. This simple-observable order now organizes the search for the theorem by finding the logic of flipping as the theorem of the reflective field. Flipping is a practical operation for finding the sense of mathematical objectivity by which it becomes a physical law of reflection. This is done by simply taking it in hand and observing its rotational order. On whatever side it falls, it unfolds the axiomatic order of the reflection while the flipping of the prism's base is a

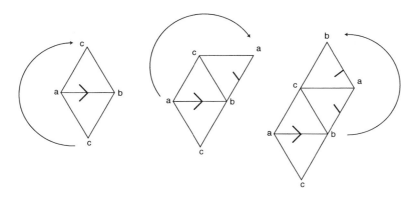

Figure 27a

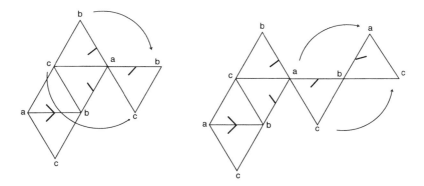

Figure 27b

premathematical, practical a priori, or a *praxiom*. By flipping the base of the prism over its three equal sides it marks the exact position of the reflected line. Every reflected triangle is the starting position for new reflections. The discovered order of flipping the simple-observable object now reveals the infinite repetitive structure of reflection as the law of the prismatic field, within the hazy degrading and limited visible field of the prism.

ANALYSIS

The ready-to-use character of the reader's own prism is made equipmentally available for self-analysis. In order to respecify the instructions of how to use the prism, the reader must align, bring his or her approach into accord, with the prism's "tacit" requirements for entering the field of use. Such requirements as to how to hold the three sided prism, where to look, how to position it vertically or horizontally, etc., are satisfied or not satisfied through its actual use. Answers to these questions are found interinstructively through the reader's self-organizing approach to the inspectable properties of the instrument ready-to-hand. For the reader, this interinstructive orientation to the instrument is the instrument ready-to-use—the prism as it is found in one's practice. At the same time, the prism, the instrument as it is found in use itself, constitutes the practitioner's own problem of how to handle it. The way in which one positions the prism and peers through it opens the possibilities as well as the limitations of what and how much one can see. Thus one's own way of using it with the discovered interinstructive order constitutes one's orientation to the instrument, which determines whether or not a transcendental code of reflection may be found. This throws one's orientation back to the "orienter" as an interstructurally organized problem at hand. It is this problem from within the prism, as my orientation to it, that the reflection, with its laws, can or cannot be discovered. Prism use is extracted from the instrument itself as a practical problem for the reader; the resolution is found in its actual manipulation. Only within this inquiry into the properties of the ready-to-use instrument can the manipulation of the instrument, as the requirement for the respecification of the instructions, be achieved. This entire prereflectively organized work will come to the reader as the already-organized work by, and as, the instrument-in-use. This, of course, presumes a "reader" who successfully masters and does not resist the instructions. We shall call this encounter with the instrument-organizing work a "Galilean moment" in the reader's biography. The reader's practical activity of reading the text, which is the true phenomenon of this book, becomes for him or her an object of practical sense-finding and serves as the requirement for reading the text.

In the course of finding a practical solution for how one is to manipulate the prism, the reader discovers that propositional knowledge and its "practical" application is of no help to him or her. Instead, the reader relies upon his or her own practice and finds in its setting the "tacit" pedagogy for an inquiry into the logic of the instrument's manipulability. Only within the field of the practice's autopedagogy can the reader find or not find the required solution. The autopedagogy is built on a scaffolding of facts (i.e., facts about the standard design of the prism, of reading, of instructional design, of apt modes of interinstructive handling and peering, none of which can be described exhaustively here or anywhere). Thus the solution to the search is internally characterized by a prereflective and organized approach to the instrument. On its way to the "marketplace" of laws, the latter is a tacit problem that mobilizes the search to find an available logic of the prism's manipulation, not as a transcendental rule or historical fact, but as a prereflective and organized outcome of the relationship between practice and object, without which "Galilean physics" would never occur.

Having demonstrated, first, the change of the subject position from the discursive to a body perspective; second, the subject-position as itself the object of a discovering inquiry; and third, that the domain of the body-instrument link is rich with a discovered intelligibility, I will now return to Galileo's pendulum in two consecutive sequences: first, Galileo's pendulum as a discursive sign, and second, Galileo's pendulum as the body-instrument link.

FIVE

The Formal Structure of Galileo's Pendulum

Historical representation of Galileo's pendulum depends on the dominant mode of rationality and its strategic suppression of the body. I have already analyzed in the second chapter Koyré's dry and strictly textual representation of Galileo's pendulum as typical for the ways in which we know Galileo's pendulum today. The main aspect of this representation is formal knowledge of a pendulum itself as a discursive sign disassociated from the *knowing* body. Schematic images of the pendulum, with mathematical formulas and tables of numbers in an abstract space that cannot be touched, made, or performed by a human hand, commonly dominate the space of scientific texts and textbooks today. Thus our knowledge gleaned from these pages remains textual, disembodied, and sexualized. In other words, our knowledge today is rooted in the Christian and Neoplatonic discourse of abstract pleasures, which were appropriated by the seventeenth-century scientist's subjectivity and disassociated from the body as a matter of personal ethics and of scientific method.

Let me proceed with the exposition of the formal structure of Galileo's pendulum. The dawn of classical physics begins with Galileo's formulation of the law of free fall. This law claims that the acceleration of a falling body (S) is constant and stands in proportion to the squares of elapsed time (T): $S = T^2$. The revolutionary significance of this law, according to Kuhn, is the radical change it caused in the perception of the world.[1] Since Aristotle, it was assumed that motion in free fall is constant and is a function of the weight of a descending object: heavier objects fall faster than lighter. Renaissance physicists revised this view but still maintained its consistency with sensual perception.[2] They claimed that motion in free fall is a function of height; namely, that objects falling from a higher point will drop with more speed than objects from a lower height.[3] The relationship between weight and motion and between space and motion is easily and intuitively comprehended. It makes sense to bodily experience that a hundred-pound lead ball will touch the ground before a feather will if both are released from the same height and at the same time, or that a dime dropped on one's head from one foot above will have a different speed than if dropped from the Sears tower. In light of this common-sense comprehension, it appears that speed is a function of weight and/or height. But Galileo's law of free fall challenges this. "Motion in space," for Galileo, may have a biased observation that hides a deeper truth about physical reality. "Motion in space" is only a symptom of an essentially temporal reality[4] in which free fall is a function of time. Time, as an abstract concept, cannot be directly experienced but only indirectly expressed by mathematics. To know and explain physical reality one therefore has to see and think in mathematical terms. Nature, according to Galileo, is written in a mathematical language.[5]

Probably the law's most far-reaching consequence for the future of "classical" and "modern" physics is the realization that time conforms to natural numbers and geometric figures.[6] Since motion, velocity, acceleration, and space are temporal categories, it follows that they too must conform to the rules of mathematics. Ironically, Galileo initially believed that nature can never conform to the ideal order of mathematics—only to find himself inventing mathematical physics.[7] Once he acknowledged the important role of mathematics in physics, the world of objects became mathematically analyzable, the physical order mathematically reconstructible, the relations of things inferred in ideal terms by axioms and theorems, and these mathematically constructed relations were quantified by measuring.[8] The realization is that space traversed through equal time units increases regularly, that is, acceleration increases constantly as do the squares of these units: A falling object descends 16 feet in the first second and 64 feet in the second second. Sixty-four is 4 x 16 feet of acceleration in the second second, meaning that the acceleration increases as does the square of the second time unit, $2^2 = 4$, and so on.[9] The rule for the acceleration of a falling body when friction is eliminated, is given in the time squared number series with its constant increase being the numerical function of time. In this sense, Galileo established the rule of motion through a mathematical relation between distance and time.[10]

A simple pendulum demonstration was designed to make an abstract geometric reality of time concrete. Galileo's pendulum occupies an important place in the pedagogy of his theory and was designed for a common-sense audience.[11] For Galileo, the pendulum was a tool of emancipation from the dogma of Aristotelian metaphysics, just as much as it was for his followers. In the simple pendulum swing, the Greek cosmos of sacred qualities collapsed and was replaced by measurable, profane quantities. The emancipational force of the pendulum came precisely because it was successful in demonstrating an abstract law of motion, the point at which the geometrical and the experiential coincide into a mathematized experience. It is for this reason that Kuhn points to Galileo's pendulum as a perception-molding device.

Through the pendulum, Galileo achieved a demonstration of the law of a free-falling body, however, the law of free fall was difficult to demonstrate or to measure directly. Galileo knew that as long as the law was not demonstrated it would remain only within the realm of mathematical speculation. Necessity being the mother of invention, the need for a convincing demonstration forced him to demonstrate it indirectly.

The purpose of the indirect demonstration was to make it possible to observe an object in free fall and to find a way to measure it accurately. Galileo's first choice in demonstrating and measuring free fall indirectly was the inclined plane. Even though the ball rolled down the smooth plane slowly enough to observe and measure its acceleration, the obtained measurement, while indicating numerical conformity with the law of natural numbers, was not precise enough to satisfy Galileo's need for a convincing demonstration.[12] Physical constraints related to the friction and flatness of the plane prevented a convincing demonstration. He needed a simple, convincing, and easily reproducible demonstration.

After the experiment with the inclined plane, which certainly convinced Galileo, if not others, that motion conforms to natural numbers, he searched for a demonstration that would be at once simple and unquestionably convincing and would demonstrate the law at a glance.[13] The pendulum certainly offered this silent "first assault" on the Aristotelian skeptics. When moved from the vertical and released from any angle, a ball hanging on a string tied to a bent nail and fixed to a wall or ceiling moves on its own and uniformly along the geometric path of an arch. With this convincing simplicity, the pendulum served as the perfect instrument for Galileo's demonstration. Galileo became "Caesar" not by inventing the pendulum, but by assigning it a theoretical purpose. The string of the pendulum, under the weight of the ball moving on its own and uniformly, keeps the path of motion equitemporal at all points while facing neither friction from rolling nor friction from an uneven or rough surface. The abstract concept of free fall is seen at a glance. The motion of the pendulum is slow enough to observe its uniformity (isochrony) yet quick enough to eliminate discrepancies, and because of its uniformity it could be used as a measuring instrument for time. Its demonstrational arrangement is geometrical.[14] The oscillation of the pendulum is within a circular structure. The nail is the center of the circle, the thread is the radius, the ball is a point on the circumference, and the lowest point is the tangent. Motion contained within this geometrical structure is analyzable according to the axiomic principles of a circle.

The simple pendulum consists, then, of a weight hung by a string, usually called a "cord," from a fixed point. The simple order of the motion produced by this arrangement is that by moving under its weight the ball must traverse along a circular path, usually called an "arc." As it moves through the arc it also moves through the vertical plane *(p)*. The "arc" has three distinct motion points: at each end is a point of descent (A, Z), while its lowest or tangential point (T) is the point of ascent. When a ball descends from the released point (A), or its angle $> \propto$, to the lowest point (T), it does not rest there but instead attains the same velocity that it would normally have in falling freely from the height from which it was released. Thus, instead of resting, it ascends upward to the other side of the "arc," reaching the same horizontal level as the initial point of release and from that point (Z) begins a new descent. From then on the motion is repeated identically. This motion is called "oscillation," "swing," or "vibration." The full swing is one in which the ball returns to the initial descending point, the double length of the arc (see figure 28).

Half of a swing is one in which the ball reaches the highest ascending point, the full size of the arc. One quarter of a swing is between the highest descending point and the lowest ascending point, one half of an arc. According to Galileo's theory, and corroborated by the naked eye, the swing is isochronous, meaning that equal distances are passed in equal times.[15] However, careful empirical and mathematical investigations since Galileo have shown that the "arc" is a much more complex temporal structure than what the eye sees and what Galileo theorized; namely, only some parts—the fastest curves—of the arc are isochronous. In other words, the isochronous property of the pendulum was largely Galileo's theoretical assumption, flawed by incomplete observation.

When the pendulum is left on its own, the ball will eventually come to rest in a vertical line. Swinging through the air, the ball experiences a slight amount of air resistance that gradually reduces the amplitude or "arc" of the pendulum and, accordingly, its velocity. Eventually the ball will stop at the tangential point. This fact alone indicates that points of descent are not symmetrically but rather asymmetrically descending. However, as long as the ball moves it still moves uniformly.

The timing of the pendulum, according to Galileo, is determined solely by the length of the cord. If one is to compare two pendula of different lengths, it becomes obvious that the pendulum with a longer cord is slower, meaning that the length of the pendulum slows the motion. On the basis of this ratio, the timing of any pendulum can be deduced from the length of its cord.

The pendulum is an example of the circular, vertical, and inclined free fall. When a ball is released from a descending point, it falls from a certain height. The pendulum is nothing but a ball rolling down an inclined plane or a vertically falling body forced to fall along a circular line. Its velocity is due to its changing levels, and not the length of its path. If the pendulum has a long cord, its path, if we consider the arc of its swing, will be flatter than if

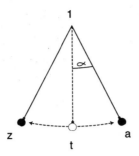

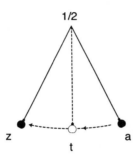

Figure 28

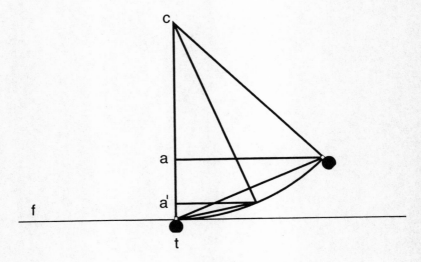

Figure 29

the pendulum were short. Therefore its velocity will be less because it has a shorter height from which to fall. Consider figure 29.

The distance CT is the radius but also the vertical plane of the falling body with a 90° angle. The distance AT is the distance of the inclined plane with a 45° angle, A' T is the distance of the inclined plane with a 25° angle.

All of the above are simple observations that, at this point, neither hold scientific significance nor coherent relationships. Galileo's mathematical theory of the pendulum explains these observations in terms of mathematical axioms and theorems which coalesce into a coherent rational whole.

THEORIES OF THE PENDULUM

Galileo was interested in the motion of the pendulum for various reasons. He used this simple instrument in connection with refuting the Aristotelian theory of impetus and the doctrine of contrary motion and with the theory of tides and nonuniformed planetary revolution. Most importantly, Galileo found a property of the pendulum—the relatively slow motion of falling bodies impeded only by air—to be very useful for both measuring and demonstrating free fall. While the pendulum was never an object of independent study by Galileo, his work with the pendulum, according to some, is sufficient enough to be regarded as a piece of research in its own right.[16] The following analysis will attempt to show Galileo's research on pendula as constituting a more or less coherent body of his science.

In reading Galileo's text, one finds different types of pendula performing different theoretical purposes. Based upon textual analysis, MacLachlan distinguishes four different discursive occasions on which the description of the pendulum's use and theoretical claims were discussed. When Galileo discusses pendula with equal lengths, he claims that a single swing of one pendulum goes as far past the perpendicular as the distance from which it was released and subsequently returns to the same height above the lowest point as that from which it was released. The pendulum reappears in relationship to Galileo's claim that every suspended ball will eventually come to rest if left to itself, and the lighter the ball, the more quickly it will rest. He discusses pendula with respect to dependence on length; the period of oscillation is shorter for shorter pendula and is proportional to the square-root of the length of the pendulum. He also claims that pendular motion is independent from amplitude, meaning that the same pendulum makes all of its oscillations in equal times regardless of the amplitude of the vibration.[17] All of these claims are related to the law of isochronism.

In order to experimentally demonstrate the above claims, Galileo used three different types of pendula (see figures 30, 31, and 32):[18] 1) nonrigid—when a ball is suspended from a string; 2) rigid—when it has a firm arc; 3) ideal—a geometric model.

Each of these different pendula furnishes different theoretical purposes, sometimes to demonstrate a law and sometimes to obtain the actual value for

Figure 30

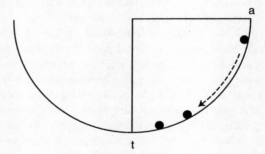

Figure 31

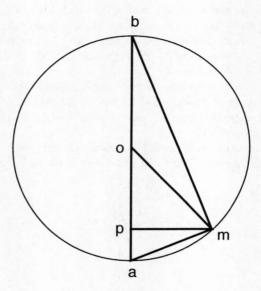

Figure 32

time. There is an intricate relationship between the type of the pendulum and its theoretical purpose.

Theories of the pendulum are appropriated to different types of pendula. There are three geometric theorems associated with the pendulum: isochronism, brachistochrone curves, and circular chords.[19] Probably the most celebrated of Galileo's claims concerning pendular motion is that of isochronism. According to this thesis, the period of oscillations is equitemporal—determined solely by the length of the pendulum—and independent of the amplitude of oscillations and of the weight of the ball. This thesis is also known as the law of length. In support of this thesis, Galileo developed two other theorems: the brachistochrone curve (i.e., the curve of quickest descent in a pendular arc), and the law of chords.

Koyré observes that Galileo's mathematical physics unified the cosmos into a coherent universal order applicable to things on earth as well as to the heavens.[20] The pendulum is a case of this unification. Galileo's pendular law of length was extended from the first law of astronomy and later served as a simplified model for his law of free fall. The first law of astronomy is formulated as: "A single movable body made to rotate by a single force will take a longer time to complete its circuit along a greater circle than along a lesser circle. . . ." In 1602, Galileo was interested in comparing the distance and speed of Saturn to other planets of the solar system. He made the following tabulation of decimal fractions as a way of comparing:

	Distance Ratio of Saturn to Planet	Speed Ratio of Planet to Saturn	Square Root of Distance Ratio
Saturn	1.00	1.00	1.00
Jupiter	1.75	1.42	1.32
Mars	6.03	2.60	2.46
Earth	9.17	3.17	3.03
Venus	12.75	3.71	3.57
Mercury	24.79	5.06	4.98

SOURCE: Stillman Drake, *Galileo at Work: His Scientific Biography* (Chicago: University of Chicago Press, 1978), 65.

Galileo found the similarity of ratios in the last two columns to be indicative of the proportion between the planetary orbital speeds and their distances from the sun.[21] Namely, that the radius of the orbital paths determines the speed of the planet.

Now, pendula with different lengths are like small planets or other bodies that move in different circles with different radii, such as a satellite or the tides. If, according to this law, the time of rotation depends on the length of the radius, then in the case of the pendulum the same relationship can be established, that is, that time is proportional to the length of the radius:

$$T \text{ (ime)} = L \text{ (ength) or, more correctly}$$
$$T = \sqrt{L}$$

By making the motion of a pendulum an example of the first law of astronomy, Galileo incorporated the pendular motion into the corpus of established physics.

In order to demonstrate the law of lengths he used the nonrigid type of pendulum.[22] His experiment is with three balls suspended from three different lengths—16"; 9"; 4"—which swing in synchrony after a certain numbers of swings—2 with the 16"; 3 with the 9"; 4 with the 4." The synchrony is hypothesized on the basis of the law of length whereby, in this case, the number of synchronic swings is deduced from the inverse square roots of the lengths of the pendula. This particular case will be discussed in length in the next chapter, but for now we are to understand that the point of this experiment is to show that the period is determined by the square root of the length.

While a nonrigid pendulum was a handy instrument to demonstrate a theory, it also suffered from limitations in actually tying the observation to a theory. Careful observation of the nonrigid pendulum shows that it is damped (see figure 33).[23] Because of the resistance of the air, the strings, in other words, bend at the end of their swings, causing successive oscillations not always along the same radius. The problem that this type of pendulum created for Galileo was that in demonstrating the law of length, the value of the length has to be both constant and established. In the case of damping, the pendulum has more than one radius.

How long, then, is a nonrigid pendulum? While the average radius can be theoretically determined, it cannot be empirically established. Since pendula damp at their extremes and thus constantly change their lengths, the successive lengths and oscillations are not constant. This is partly because of the impediment of air on the string but also because of the different weights. Heavier balls make strings more resistant to air impediment and damping and so different weights produce different damping and conse-

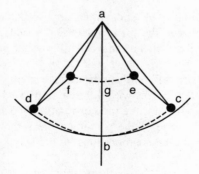

Figure 33

quentially different amplitudes. A careful empirical observation of Galileo's pendulum is not always consistent with the premise that timing is independent of weight.[24]

This problem is clearly observable in the case of a pendulum with different weights. The claim that the period is independent of the amplitude of oscillations may be demonstrated by the nonrigid pendulum. For example, two equally weighted balls with different amplitudes closely approximate the theoretical claim. However, when the weight changes, for example, when you use one steel and one cork ball, the discrepancy in amplitudes is significant and the results are inconsistent with the original theoretical claim.[25]

In the demonstration of the law of length using the nonrigid pendulum, Galileo did not succeed in proving the isochrony of the pendulum. All of his demonstrations proved only the synchrony of pendula with the same lengths, or of pendula in which time stands in an inverse ratio to their lengths. The nonrigid pendula also fail to show that all simultaneous or nonsimultaneous oscillations of the pendulum have equal amplitude.[26]

To come closer to his theoretical conditions of pendular motion in an effort to demonstrate the law of length more convincingly, Galileo used the hoop as a rigid pendulum. This type of pendulum is less handy, but supposedly it eliminates damping and allows for equal timing.[27] However, the rigid pendulum carries its own set of problems. One of Galileo's friends complained about not getting equal time from two bronze balls on the rigid hoop. Galileo blamed this on the concave form of the hoop. The rigid pendulum is not always as circular as the ideal pendulum should be.

THE IDEAL PENDULUM AND ITS THEOREMS

While the nonrigid and the rigid pendula are easy instruments to use, they are imperfect and can only approximate an "ideal" demonstration. This explains why Galileo uses expressions such as "almost" or "close enough" when working with them.[28] To achieve clarity of demonstration, Galileo used an ideal geometric pendulum as the best tool to demonstrate the law under ideal conditions. He used it to demonstrate geometrically what he could not demonstrate empirically. The ideal pendulum is one where the pendular process is reduced to its fundamental ideal elements. What serves as the thread in the nonrigid pendulum is the radius of the circle in the ideal pendulum and what serves as the ball is a point on the circumference. With the ideal pendulum, the damping and curving of the strings and the concavity of the hoop are eliminated. Although these theorems were significant for Galileo's thesis on isochronism, they were never proven through demonstration. It took Huygens to accomplish that.

The Theorem of Isochrony

The theorem of isochrony was formulated in October of 1602 in a letter from Galileo to his friend Guidobaldo. In this letter, he argues that speed is constant regardless of the amplitude of the arc's descent. Stillman Drake explains this claim as follows (see figure 34):

> It is assumed that two different speeds along AB and AC, each being regarded as uniform and dependent only on slope, have the ratio of distances AC and AB, respectively; it is so to be shown that the times along AE and AF are identical. Since AD is the mean proportional between AE and AB, as also between AF and AC, we have AE : AD : : AD : AB and AG : AD : : AD : AC. By division, AE : AF : : AC : AB. Now, AC : AB was taken as the ratio of speed along AB to speed along AC, and is seen to equal AE : AF. But when speeds are proportional to distances traversed, the times are necessarily equal; hence AE and AF are traversed in the same time.[29]

Accordingly, speeds S(AB) and S(AC) are proportional to their distances and traversed times are necessarily equal. Times T(ED) and T(FD) are necessarily

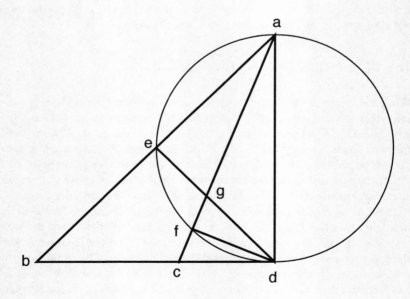

Figure 34

equal. This supports the claim that amplitude (arc ED and FD) does not determine the time of motion along the pendular arcs of the same circumference but only determines speed, since they are proportional to their angles.

The Theorem of Chords

The theorem of isochrony and experimentation with the pendulum led Galileo to the next two theorems. First is the theorem of chords, in which Galileo states: "If from the highest or lowest point in a vertical circle there be drawn any inclined planes meeting the circumference, the time of descent along these chords are each equal to the other."[30] Galileo writes (see figure 35):

> Let BA be the diameter of circle BDA erect to the horizontal, and from point A out to the circumference draw any lines AF, AE, AD, and AC. I show that equal moveables fall in equal times, whether through the vertical BA or through the inclined planes along lines CA, DA, EA, and FA. Thus leaving at the same moment from points B, C, D, E, and F, they arrive at the same moment at terminus A; and line FA may be as small as you wish.[31]

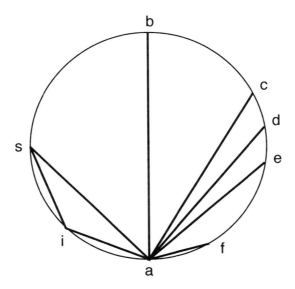

Figure 35

The Theorem of the Brachistochrone Curve

In the theorem of the brachistochrone curve, Galileo states: "If from the lowest point of a vertical circle, a chord is drawn subtending an arc not greater than a quadrant, and if from the two ends of this chord two other chords be drawn to any point on the arc, the time of descent along the two latter chords will be shorter than along the first."[32] He writes (see figure 35):

> And perhaps even more surprising will this, also demonstrated by me, appear: That line SA being not greater than the chord of quadrant, and lines SI and IA being any whatever, the same moveable leaving from S will make its journey SIA more swiftly than just the trip IA, starting from I.[33]

Galileo claims that the times of descent are shorter along the broken paths than along the direct chords. That is, the more the broken paths look like a circle the shorter the time required for descent from C to A or S to A. "From the proceeding it is possible to infer that the path of quickest descent from one point to another is not the shortest path, namely, a straight line, but the arc of a circle."[34] However, such arcs are passed in equal time only when they are very small, meaning that a general proof for all arcs was not possible.

The theoretical coherency of these three theorems is made possible by the structure of the ideal pendulum. A moving point follows a circular arc just as the ball of a rigid pendulum does, but avoids the imperfection of the rigid surface and the damping of the nonrigid pendulum. According to the theorem, when the ball descends along the ideal arc the time is minimal. Galileo inferred that since the times of descents from rest along the chords of the circle are all equal, and since the times of descents from rest along the arcs corresponding to these chords are all minimal, the times of descent from rest along these arcs of different lengths ought to be all equal—isochronous.[35]

These theorems and the pendulum thesis of isochrony relate in the following way. Galileo treats chords as inclined planes, along whose heights the pendulum's ball descends and which, at the same time, mark the length of the ball's amplitude. The descent of the pendulum's ball is along the arc. Both, along the plane and the arc, are examples of motion due to gravity force. Both are examples of free-falling bodies. It is evident by now that, for Galileo, different types of pendula served different types of theoretical purposes. His ideal and rigid pendula were theoretical devices for rendering hypotheses while his nonrigid pendula were instruments that demonstrated empirical approximations of the theorems or the indirect measurement of time.

SIX

The Respecification of Galileo's Pendulum

The lead balls used here to construct Galileo's pendulum were obtained from *Nasco Science '96: Understanding Through Hands-On Science* catalogue. When I received the catologue, I realized that finding Galileo's pendulum in a book of hundreds of pages is in itself a *method* of discovery. I had to engage in holding and looking at the catalogue, and flipping the pages comparing texts; in other words, figuring out the logic that would show me the path to pendular balls. This path had the following sequential order. On the third page of the catalogue, "Quick Index," a general index provided in alphabetical order that covers the entire inventory according to various interest groups paired with the page locations within the catalogue. So for example, to find Galileo's pendulum the following order should be followed. Under the letter "P" in the "Index," one finds the category of "Physical Science Teaching Aids" and looks for "Pendulum." There is "Pendulum Balls" p., 322. On that page under the title "Nature of Motion" the list of various pendula can be found. The first are "pendulum and collision balls," followed by "pendulum clamp" and Foucault's pendulum. Their specifications and prices are included, too. For instance, the lead balls used in my project, 1" wide, cost $4.70. The other way to find "pendulum" is from the front "Quick Index" under "Physical Science Apparatus" (pages listed above).

Although the catalogue may generate and satisfy the need for finding a pendulum, it does so only as a fetish of science. The catalogue displays pendular balls in various sizes, materials, forms, and prices to accommodate various educational needs, such as their use as tools for recovering a discursive history of Galileo's science or, "new mechanics." In addition to their historic place, the balls also encompass organizational features of the catalogue, the *place* and the *path* of finding them. However, the catalogue does not supply the reader with the "handmaid knowledge" needed to make the instrument actually perform their designated functions. The catalogue only presupposes that when the instrument comes out of the package it will do what it says, i.e., that the pendulum will demonstrate the law of the period of a pendulum, defined mathematically as 2 times pi, divided by the square root of gravitational constant (g) and multiplied by the square root of the length of the pendulum (l) or,

$$T = 2\pi \sqrt{\tfrac{l}{g}}$$

But the relationship between the pendular balls from the catalogue (plus the mathematical formula from the textbooks) and the user is, as the previous chapter has shown, extracatalogic and has no prespecified path of finding the phenomenon. The balls need to be assembled into a pendulum and properly used in order for them to be what the catalogue claims they are. When they arrive into the hands of their user, their *place* in the catalogue and the actual *body-site* of their assemblage must negotiate their name. The missing body-knowledge from the catalogue is a consequence of the sexual history of science as a history of a perversion, of a systematic exclusion of the body from truth. Presenting pendular balls without their link to the body-work stands as a pure fetish of formal knowledge. In their glitzy, perfect and high-tech shapes, scientific instruments discern themselves as "soft-porn" signs of their history.

In opposition to the "catalogue knowledge" mostly represented by Koyré some historians of science were determined to uncover the very discovering features of Galileo's pendulum. For that project they had to employ their bodies to fill in the blanks. After their reconstruction of Galileo's experiment, Stillman Drake and John MacLachlan reported that they developed "a theory of Galileo's procedures" that could be tested like "any physical theory, namely by actual experiment."[1] Some of this "theory," according to Thomas Settle and his own discoveries of Galileo's inclined plane experiments, includes the experimenter and his or her *feel* for the equipment as an *integral part* of the measuring procedure. Settle emphasizes that the experimenter must consciously *train* his or her own reactions to the equipment. Before any measurement, the experimenter must be allowed to practice a few runs to *warm up*. Only after many trial runs and the *routinization* of the procedures can measurement of the pendulum's "pulse beat" take place.[2] In different ways, then, Settle, Drake, and MacLachlan discovered what Koyré excluded from his history of Galileo's science, the bodily dimension of Galileo's instruments and experimental procedures tacitly built into Galileo's formal knowledge.

Unlike Drake, Settle, and MacLachlan, however, ethnomethodologist Eric Livingston theorized this discovered pairing complexity of Galileo's experimental procedures, and made them explicit. He claims that Galileo's experiments consist of two segments: a) "propositional space" and b) "domain of worldly theory." While "propositional space" is the representational vortex of formal propositions, mathematical deductions, and explanations of "worldly phenomena," the "domain of worldly theory" consists of various objects, physical properties and procedures all aimed to exhibit the claims of "propositional space." Between these two opposing poles or "Livingston's pain," operates "scientific method" which, with its standard procedural discourse and operations, ensures "the correspondence between the propositional space and worldly phenomena."[3] Thus, according to Livingston, the structural problem of representing Galileo's science seems to be reducible to the following formulation. If the first part of the pair, "propositional space," contains the theorem account, then the theorem account must be the precise description of the second part of the pair, that is, of the lived work of proving. The correspondence between the transcendental claim and the lived work of proving, obliterates the very lived work. Ethnomethodology, by emphasizing scientific practices, makes the obliterated work visible, which I am about to address in the remaining part of this chapter.

In this chapter I shall first discuss the "pendulum" as a narrative structure and the ways in which the body "extracts" Galileo's pendulum from his text. As the analysis below will demonstrate, this path begins with positioning the body towards Galileo's text and the bodily production of meaning pertaining to the work of making the pendulum.

Let me start with the textual structure of the pendulum and then move to its link with the body. The pendulum has a textual structure as well as an operational structure. There are two different textual uses of the pendulum by Galileo: one in the books, for didactic purposes, and the other in experimental notes, for discovering purposes. However, this distinction is not strict, since the text states the a priori structure as well as instructs the operator. The pendulum as a textual structure is part of Galileo's scientific narrative. The pendulum is introduced in the text on various occasions but, each time, it is used as a didactic device for Galileo's a priori claims. In the text he distinguishes different types of pendula described for different textual purposes. For instance, if Galileo wanted to convince his friend that the inconclusive experimental findings with the rigid pendulum was due to the surface, he would utilize an ideal pendulum to mathematically prove his claim. If, on the other hand, he wanted to prove the square root of the length theorem, he would introduce the nonrigid pendulum.[4] What is common, however, on all of these occasions is a sense of demonstrational optimism due to Galileo's certainty of his a priori claims.

The pendulum as a text appears in Galileo's experimental notes in the form of a sketch, either as a mathematical structure or as a real model.[5] The real pendulum appears to be used as a measuring instrument for time, a "time keeper," while the ideal pendulum serves to determine mathematically which curves of the swing to use, how to find the time of the free fall through the radius of the pendulum, etc.[6] Usually the sketch is surrounded with numbers produced through some kind of calculation or recorded measurements. From Galileo's notes, it becomes clear that the pendulum as a demonstrational tool was used in the textual narrative, while the pendulum as a discovering tool was used in his experimental notes.

The following is an example of how the text is instructive in discovering the unexplained procedures for the pendulum as an operation. Here is the a priori structure of the pendulum.

A priori

> As to the ratio of times of oscillation of bodies hanging from strings of different lengths, those times are as the square roots of the string lengths; or we should say that the lengths are as the double ratios, or square, of the times.[7]

The theorem is formalized as:

$$T = \text{Time}$$
$$L = \text{Length}$$

$$T = \sqrt{L} \text{ or } L = T^2$$

Deduction/Operationalization

In continuation of the above quotation:

> Thus if, for example, you want the time of oscillation of one pendulum to be double the time of another, the length of its string must be four times that of the other. . . . It follows from this that the lengths of the strings have to one another the [inverse] ratio of the squares of the number of vibrations made in given time.[8]

On the basis of the above claim as a priori we can deduce the order of lengths required to produce "double time" synchrony among the swinging pendula. Note how Galileo's deductions already include a description of the operationalization of the a priori claim. Namely, cutting and tying strings must be in the 1 : 4 proportion; one needs to utilize these skills, yet a directive has to be followed at the point of cutting and tying or, let me say, in the *montage* of lengths, in order for these skills to turn the available materials of nails, strings, lead balls, ceilings, etc., into the proposed synchronization of time. Here is where the conducting of an experiment begins involving, in Livingston's parlance, "the cultivation of the pairing of equipment and practice . . . so as to provide a stable background against which the experimental demonstration is exhibited."[9]

OPERATIONAL CLUES

To account for this theorem, the following specifications of the pendulum were given by Galileo in order to construct it and prove the theorem by its use:

> Hang lead balls, or similar heavy bodies, from three threads of different lengths, so that in the time that the longest makes two oscillations, the shortest makes four and the other makes three. This will happen when the longest contains sixteen spans, or other units, of which the middle [length] contains nine, and the smallest, four. Removing all these from the vertical and then releasing them, an interesting interlacing of the threads will be seen, with varied meetings such that at every fourth oscillation of the longest, all three arrive unitedly at the same terminus; and from this they depart, repeating again the same period.[10]

Note that Galileo did not specify *how* to build the pendulum, *how* to tie a knot, *how* to make a swing, *how* and from where to look at the swings, or *how* many people are needed to do the witnessing. This is to mention just some of the practical contingencies that come with the theorem and the pendulum with only very sketchy directives.

It should be clear at this point that I take Galileo's science to be a complex of exemplary situations with the required tacit knowledge of and skills for managing those situations in order to achieve a particular conceptual awareness. Being a complex of exemplary situations, "Galilean science" is not reducible to only its textual form. A feature of "Galilean science" is that it consists, as we have seen, of two constitutive components: 1) the text as a formal structure of knowledge, and 2) the actual work of doing science (including reading the text). Both are necessary for an adequate understanding. For example, the law of pendular isochronism, discovered by Galileo, is textually stated as: the pendulum ball passes equal distances in equal times; taxonomically it is represented as: $T1 = T2 = T3 = \ldots Tn$. Now, Galileo's physics is revolutionary because it introduces the experiment and the empirical demonstration of its theoretical claims. Thus, an intricate part of Galileo's physics is to convey to the reader his conception of motion as the reader's perception of it. Galileo offers exemplary situations in which the reader sees the pendulum swinging uniformly, independent of amplitude or weight; Galileo's text *makes* nature conform to the reader's vision. Sometimes this situation is a geometrical proof of his claim; sometimes it is a reference to a common experience that many people of practice encounter and are either aware of or not; sometimes, and this is the revolutionary aspect of his physics in comparison to Aristotelian physics, he gives instructions on how to perform the experiment and experience the phenomenon empirically; and sometimes he gives a report on his own experiment.

The claim that the pendulum is isochronous is something that has to be demonstrated, meaning that the formal expression $T1 = T2 = T3 = \ldots Tn$ has to be exemplified by the actual motion of a pendulum. At this point Galileo's text extends into the nontextual field of tacit knowledge and skills for producing exemplary situations of isochronic motion. To actually achieve this, Galileo offers instructions on how to make a pendulum: find a string and a ball, tie it to a bent nail fastened to the ceiling, and swing it along the vertical plane. At this point, the text reaches its pedagogical limit. At one point, as Wittgenstein says, instructions come to an end. The task of following instructions is not in the text any longer but, rather, in the exemplary situation in the pedagogy of the body-pendulum link.

Following the latter rules Galileo's pendulum assumes the following representational order:

Text	Exemplary situation
theorem:	
$T1 = T2 = T3 = \ldots Tn,$	
	$\longrightarrow$
instructions:	
tie a string to a ball ...	
Fasten a bent nail to the ceiling ...	
have three strings ...	
cut ...	

The fact that the right side of the page is empty is not an oversight but a clarification of the pedagogical structure of Galileo's science. With the text, the closest we can come to the pendulum is to *imagine* its mechanical structure, but only the body, the skills and the materials can make pendulum. At the stage of reading the text, the embodied part remains as an unknown variable. The blank surface of the page refers to the "worldly domain" of the pendulum, to the ending of discursive rules and the beginning of practical rules. Here is where the pendulum explodes its textual form. Certainly we hold the text not only as a body of propositional knowledge but also as a clue for how to operationalize textual claims. The pairing of tools, practices and textual claims shifts the referent from pure discourse to the body-instrument links. The exemplary situation of making the pendulum integrates the text as only one pedagogic component, not as a totality of Galileo's knowledge, but as a pairing component. Under these circumstances, the meaning of the scientific text is not assured by mathematics but by the lived work of making the pendulum. While the mathematical meaning comes to the scientific text a priori, the practical situation brings meaning to the text a posteriori. Having said this, I hope to have created a sense of understanding for keeping the blank side of the above scheme. What belongs there comes a posteriori. In this sense, the formal presentation of Galileo's science in the science textbooks is always built upon an absence whose content is concealed in the laboratory work. Galileo's science seeps deeply into the everyday practical world, and only in the action of doing science can the completeness of Galileo's science be grasped.

"Galilean science" is reproduced as it was produced, as an actual work of reading and writing, making and using the instruments, and understanding relations through exemplary situations. The textual description of Galileo's pendulum is a formal account of what the informal structure of and conceptual achievement in the exemplary situation is. The informal and tacit structure of exemplary situations is something that cannot be reprinted as its formal structure but has to be encountered first-hand and discovered.

DESCRIPTION OF THE PENDULUM

When Galileo proposed the specifications for the pendulum (or any other instrument), he unintentionally left a set of practical contingencies for the practitioners to find and resolve according to the specific local conditions of their work. In short, the structures and their descriptions of the discovery of motions of hanging bodies are available only where the discovery is reproduced—in the pendulum and in the demonstration.[11] The respecification of Galilean practice requires the respecification of the pendulum and the demonstration in order to describe the practical contingencies through which the proof is achieved. This operation begins with a change in the textual purpose at hand; the text must be read instructively. In the following description, the instructive aspect is underlined and selected from the rest as practically relevant. It is treated as a practical clue for the operationalization of the description.

> Hang lead balls, or similar heavy bodies, from three threads of different lengths, so that in the time that *the longest makes two oscillations,*

the shortest makes four and the other makes three. This will happen when the <u>longest</u> contains <u>sixteen spans,</u> or other units, of which the <u>middle</u> [length] contains <u>nine</u>, and the <u>smallest</u>, four. <u>Removing</u> all these from the <u>vertical</u> and then <u>releasing</u> them, *an interesting interlacing of the threads will be seen, with varied meetings such that at every fourth oscillation of the longest, all three arrive unitedly at the same terminus; and from this they depart, repeating again the same period.*[12] (Italics and underlining mine)

The underlined words are ones that have the instructive character of "what" and "how" to assemble and to demonstrate in order to see what is only imagined by the work of the text (italics). While the italicized part of the text brings out the contours of the pendulum, the reading of the underlined part of the text emphasizes the descriptions of operations. By switching the purpose of reading, the textual signification changes. This suggests that reading, rather than a text, is the first part of the pair. Reading is a project-at-hand with a purpose, which, depending on what the purpose is, in turn singles out which textual details are more relevant and what they mean. Descriptions of operations, then, are not actual operations, they are only given in a virtual form. The italicized parts of the text describe the visual structure of the demonstration. In reading it, one *sees* the synchrony while in reading the underlined sections, one is instructed on how to make synchrony *be* seen.

RESPECIFYING GALILEO'S PENDULUM

Textual clues provide the first step towards the operationalization of the pendular theory. Rebuilding Galileo's pendulum and demonstrating the corresponding phenomenon of motion revealed to me what the assumed discovering structures of the laws of isochronism and free fall are, and what is the missing part of Galileo's physics. The first step for recovering the missing part was actually to build the pendulum from the text; the second was to make it work as visualized; and the third was to describe the missing part.

The next step was to visualize the instrument and the demonstration. Three pendular balls were hanging on three different length strings from a horizontal pole, swinging in occasional synchronization. Then I proceeded to construct the instrument as imagined. I began by finding lead balls. Through a catalogue, I ordered three lead balls of the same size (3/4") with the shape shown in figure 36. My next step was to tie the balls to strings of specified lengths and onto the "hanger"—a pole to which the string is fixed. Since Galileo did not specify how to do this, I assumed that any string would do and any object with a horizontal position from which stringed balls could hang

Figure 36

would serve as a "hanger." So I looked around and found in my son's fishing box a roll of fishing line that I could use for string. My table lamp appeared to suffice as a horizontal hanging pole for the hanger. I knew that fishing line will stretch if a heavy object hangs from it, but since the balls are of the same weight the ratio of the stretching would remain the same. I did not speculate that maybe there could be forces acting on the string other than weight that could alter the stretching to an extent that would not correspond to the square of their lengths. I realized that Galileo's unspecified procedures left room for me as the demonstrator to engage in making a common-sense physics by selecting materials and assembling them according to the presupposed forces and achieving a sense that this "will do" for the purpose I have in mind.

What Galileo did not say, and what was left to me to discover, was how to tie the string to the balls and to the "hanger." The balls had a place for tying but I did not have instructions on how to tie. I was not clear if the knot itself should or should not be included in the specified length of the string. My conclusion was that the knot itself should be part of the length since it swings with the rest of the line. Therefore, I assume the knot should be understood as the beginning of the specific length (only later to learn that the center of the ball should have been the beginning or end of the length). So, by looking at this part of the ball and tying the line through it, I assumed that the lengths of the lines (4":7 $1/9$":16") are to lie between the knot at the ball and the hanging pole. It became clear to me that if I made a knot on one side of the string (the ball or the hanger), the other knot had to be made in a way that ends exactly at 16," 7 $1/9$" or 4" from the other knot (see figure 37).

What is presumed here is that I, as the demonstrator, have the skill to tie a knot at the predetermined point (x) on the string and that the replicator's fingers already "know" how much string it takes to tie a knot around

Figure 37

this particular ball and this particular hanging pole, all of which I had never done before in this manner. The part that I could not master was tying the string at the marked point (4"; 7 ¹/₉"; 16") to the lamppole without creating an extra space between the knot and the lamppole.

The first problem, of exactly replicating the string lengths, emerged in the space between the ideal position of the knot and the achieved one. The second problem was that this uncontrolled space is by no means standardized, meaning that every knot might have a different space between the actual and the ideal position of the knot depending upon the material and my manipulative skills. A lack of skill in tying the ideal knot did not allow for exact specifications. This in turn created a lot of "noise"; the swings were not quite as synchronic as they should have been, and showed only repetitive tendencies toward the ratio rather than exact, synchronized meetings at the terminal. Also, they would not swing neatly on the plane, but after a few minutes would swing erratically and interfere with the routes of the other. However, after doing my best (and not without frustration), I settled not for the exact measures but for the best that I could do with the available objects at that time and for the purpose of writing this text.[13] This led me to realize that my knots were achievements of my skills (i.e., that the string I had and the pole with all of its unique features organized my skillful work of knotting in such a way that the assembling order of the pendulum describes adequately the skills incorporated in it). By this, I mean that by virtue of my practical skills the instrument in front of me—the strings, the balls, the lengths, and the distinct positions—embodied Galileo's law as the operational order of it.

THE INTERSUBJECTIVE ORDER OF GALILEO'S PERFORMATIVE MATHEMATICS

Building Galileo's pendulum is not simply a matter of putting different parts of the pendulum together. Doing it self-reflectively is above all the way to discover the practical contingencies of the workings of Galileo's pendulum. I will briefly outline two discovered praxeological-related regions of Galileo's pendulum: one is the region of *observation* and the other of *operation*. In the following section of this paper, I analyze in detail the praxeological structures.

Operational Region

The operational region of the instrument is one in which the demonstrator approaches the instrument for the purpose of producing a witnessable demonstration. Within the overall division of standpoints of the pendulum this region is different in standpoint from the witnessing region of the demonstration in one important respect: it requires unique skills for manipulating the pendulum's delivery of the phenomenon of synchronized motion.

Contingency 1—How to approach the pendulum

The first discovery of the *Lebenswelt* structure of this region relates to the formally unspecified and yet required intuitive body orientation to the operation of the instrument. For the instrument to be used, the demonstrator has to incorporate the instrument into his or her collection of habitual practices. In my case, I discovered that I incorporated the pendulum from the right side, facing it as my prereflective selection. There were no instructions in Galileo's text as to the side from which the demonstrator should approach the instrument. The pendulum's entrance is surrounded by the intuitive field of the practitioner's body. From a formal point of view, it is not relevant to the purpose of the demonstration, but it is praxiomic in terms of operation (i.e., in order for the pendulum to be used it has to be approached somehow). This problem was left for the body to resolve intuitively. The body moved prereflectively into the field of the instrument, thereby establishing one's habitual relation to it, just as it did Galileo's.

Contingency 2—Two hands, three balls

Starting to perform the demonstration led me to another discovery. At the point of starting the demonstration I faced a seemingly trivial problem: how to hold three balls on the same plane with only two hands.

This issue was actually among the most significant I encountered. It helped to bring to the surface two important inferences. First, there is a unique logic to the field of the invented pendulum that sets up the practical contingencies for its operation beyond the time of its origin. Renaissance scientists, as well as undergraduate students in American colleges face the same practical problem: How to hold three balls with two hands? It is an anthropological dilemma. Second, this anthropological constraint requires a tacit inquiry into the unique order of the three-balled instrument in order to find a two-handed solution.

Contingency 3—How to hold the balls

Holding the balls had its own practical requirements. I felt the weight of the lead balls hanging on the rod on my palm. While resting on my palm, they would often escape and swing erratically. Palms are not smooth surfaces on which three balls can be easily placed, as if on an inclined plane (sometimes I would use a large hard cover book as a plane instead of my palms). Yet, there is no way but to use the palms. I had to learn how to contain the balls in my palm, on the same inclined plane, and to release them simultaneously at will. It required time to familiarize my hands with these contingencies. Over time I felt the balls intuitively, and I learned how to make them swing in a concerted fashion. The lead balls were not a philosophical object for me, a thing in the Kantian sense of a thing-unto-itself, but balls as an object organizing my practice; they are balls that swing, that escape my control, but also balls that can be tamed to swing synchronically with other balls. In other words, the observable mathematical ratio, an order logically independent of the physical contingencies when applied in Galilean physics, appears to be the rule to which the motion of hanging bodies conforms when swings are smooth and synchronized by demonstrational skills. This led me to the next discovery.

Contingency 4—Making the inclined plane

Setting the balls up to swing in synchrony is the next task of the operational region. To achieve this, the body is required to attune itself to the order of the three lead balls for the purpose of a successful demonstration. As I was having difficulties in producing the synchronized motion of three balls with only two hands, I realized a very important distinction between different starts in the demonstration. I learned that the start, the moment of attunement of the body to the balls, is a prereflective practice. For example, every start is different from the previous one. However, each can be typified as successful or unsuccessful depending upon the degree of the body's attunement to the balls. The result is either erratic or synchronized swings. A successful start consists of two achieved components. First, all three balls must be positioned on the same inclined plane (my palms or a book cover). The second component is that the balls have to be released simultaneously from the same inclined plane. If one or the other component is not achieved, the swings will not demonstrate the claimed synchrony and ratio.

In achieving a successful start, I learned that while the motion of a released ball is a natural fact, synchrony is not. Synchrony is a temporal achievement of the object-relational-skills, which themselves are unexplicated achievements of the very same discovering practice of Galileo's physics. Discovery through an achieved synchrony is a part of the *Lebenswelt* structure of discovery in Galileo's pendulum. The pendulum incorporates these skills into the phenomenon of motion, making it appear as a natural facticity and not a practical accomplishment of the very same skills. However, to conceal the workings of the hand in the operational region is essential for the mathematics to emerge in the witnessable region. Thus the mathematical account in the witnessing region is praxeologically tied to the skills from the operational region. One might argue that the Greek mathematics applied by Galileo is operating here not as a form of "rational memory," to use Bachelard's term, but as a set of embodied skills related to the project.[14] Not only are the skills of building the pendulum a prerequisite for mathematics to be witnessed, but also, mathematics presumes a successful and pre-reflective inquiry into the practical logic of the operational region as the achieved and yet unexplicated "order" of the phenomenon accounted for as a mathematical order. Witnessing the mathematical law in smooth synchronic swings is certainly a great achievement of the practical grace so unique to the Renaissance craftsmen. But it is also a tacit criterion for the competent construction of the demonstrational instrument and its use.

That any physicist was, is, and will be facing the above-described contingencies under ordinary circumstances is unique and anthropologically conditioned for this specific pendulum. Galileo ignores the fact that his instruments require tacit practices when he neglects to give us a cue and leaves it as a tacit problem for the physicist to discover. But in so doing, he created physics as an explicated order of the very discovering science of physics. Without solving this initial silent problem there would not be a dissemination of Galileo's demonstrational practices.[15] This silent problem requires an on-the-spot discovery of the "how of its origin," not, I should emphasize, in terms of the Husserlian "how" of the "transcendental ego's" intentionality, but in terms of the instructively respecifiable "how" of practical detail.[16] As the demonstrational instrument is assembled and used, "Galilean physics" as an explicational method is extracted from the instrument's use as a practical inventory of the order of the instrument. It is inseparable from the acquired skills in a way that even the historic Galileo himself would have had to acquire if facing the same practical contingencies. This is what was meant when I proposed that specific skills are required in order to enter the field of demonstration.

So, I finally solved the problem of the balls by resting the ball of the longest string in the palm of my left hand and the balls with the middle length and shortest strings in the palm of my right hand.

Contingency 5—How to count swings

In holding and releasing the balls, I gradually achieved more and more synchronic swings. As the synchrony became more distinct it became clear that another problem was emerging: How to account for three separate movements of the balls at the same time? I discovered that the three separate motions can be observed by a single person only as one intricate *Gestalt*. Since the three different motions have to be counted separately and simultaneously, this cannot be done from a single standpoint (which is not to be confused with a single individual or a personal view). Here is another anthropological contingency that requires a division of observational labor. Counting must be done *collaboratively* in order for the ratio to be observable. Otherwise, swings remain in every instance an "occult" object. Here is an important point regarding the order of witnessing in Galileo's demonstration. To make what is "occult" visible requires the viewer to do the counting of the individual swings, either singly, one after another, or, if it is done by three viewers, simultaneously. For one to witness the ratio, one has to either count every single ball's swing in order to compare it with the rest of the swings or to combine one's own account with the account of others. Only in calculating and comparing them with other swings can the mathematical ratio of the swings be demonstrated. This means that only by doing separate or combined counts and comparing them can one "witness" the ratio.

THE WITNESSING REGION OF GALILEO'S PENDULUM: MAKING GALILEO'S STANDPOINT INTERCHANGEABLE

In contrast to a scholastic meditative physics, "Galilean physics" is the public activity of demonstrating and witnessing the workings of scientific instruments. While scholastic physics presupposes a personal faith in the order of the universe according to a given divine a priori, Galilean physics presupposes an interchangeable standpoint between the discoverer and the public as an intersubjective ground of agreement in phenomenon.[17] Because of its intersubjective character, this new kind of physics requires instruments for achieving the reciprocity of standpoints.

Contingency 1—More than a single standpoint in accounting for the phenomenon

With the invention of the pendulum (and herein lies the important praxeological function of scientific instruments), Galileo's initial standpoint of looking at the object swinging and seeing it as an isochronic motion becomes, in the workings of the pendulum, an instructively reproducible standpoint. Thus the witnessing of the instrument's demonstration is a way to convince the skeptics of the initial standpoint of the discoverer.[18] The pendulum was so constructed that its requirement for intersubjective witnessing was incorporated into the order of its assemblage (like building the jury bench in the court room).[19]

From Galileo's story we also learn that Galileo's pendulum needed at least two standpoints in order to account for the demonstrated phenomenon. Through a collaborated witnessing, one might say, a theoretical claim became empirically observable for Galileo and his friend. In fact, as we will see, a part of the discovery in respecifying Galileo's pendulum is that the mathematical ratio of the pendulum's swings cannot be observed from a single vantage-point. Only through the "oscilloscopic division of labor" can the mathematical order shine through the pendulum and be verified as an intersubjective achievement. I will discuss this issue in detail later.

While being a publicly oriented work of discovery, Galileo's pendulum does not "radiate" the order of the phenomena on all possible sides and for all times. Only under certain angles of viewing distances, or time spans can one observe a more or less noticeable tendency of the assembled objects to conform to his theoretical claims. The reproducibility of Galileo's demonstrations is always a locally witnessed order. The following are some of these local contingencies.

Contingency 2—The operational region must be perpendicular to the witnessing region

The witnessing region stands perpendicular to the operational region. While the operator is facing the balls swinging one next to the other, the witness sees them swinging one behind the other. As we have seen from the analysis above, what is going to be seen by the witness in part depends upon the skills in the operational region. With the skillful operation of the pendulum, three lead balls with three different lengths ought to show synchronic movements after a certain number of swings. The meeting point of synchrony is the vertical line—the lamppost itself—that the balls conform to when at rest. The order of the synchronic arrival of the three balls at the terminal plane when the balls are released simultaneously from the same inclined plane is after two full swings of the longest, three of the middle, and four of the shortest. The ratio of the swings 2:3:4 is the result of the inverse square of their lengths. So, a 16" long string and a 4" long string determine each other's swings as the inverse ratio of their length squared (i.e., the square of 16 is 4, thus four swings for the 4" string; the square of 4 is 2, thus 2 swings for the 16" string).[20] This synchrony demonstrates the conformity of the motion to the mathematical rule.

Contingency 3—Embodied geometry and how to find the "inverse square of time"

The following geometric order is incorporated into the pendulum with one demonstrational goal—to show the "inverse square of time and length." The vertical plane is perpendicular to the planes of the balls and is extended into the field of witnessing. Each ball's swing has its plane perpendicular to the terminal plane and parallel to the planes of the other two balls' swings. So the pendulum presumes the following planes (see figure 38): the inclined plane *a* for the start; the three vertical and parallel planes of the balls *b; c; d;* and the vertical/terminal plane for the display of the ratio *f*. The internal geometric relations between these planes assemble all parts of the pendulum (the balls, lamp, strings), the scientist's skills, the observational field, and all of the parts of the performance and demonstration into one witnessable unit: the temporal ratio of motion. While three parallel planes display a uniform motion of a single ball, the terminal plane standing perpendicularly to them organizes the vantage point from which the balls are seen in a row as a temporal cohort. Parallel planes display single motions; the terminal plane invites their analysis. The internal logic of the pendulum's planes transforms the simple motions produced within the operational domain into a theoretically structured temporal cohort. This achievement of the pendulum's internal order is exhibited in the seen synchrony of motions in the designated meeting point. While the motion is the "natural" aspect of the pendulum's operation, the synchrony and the witnessed ratio of the swings belong to the aspect of the pendulum's demonstrational lucidity—to its internal organization for "finding the inverse square of time and length." Because the point of synchrony is in the terminal plane, the angle of viewing must be perpendicular to it. In other words, the internal order of the performative geometry of the pendulum and the transfigurational character of the terminal/analytical plane is extended into the field of public witnessing. For the viewers to see the synchrony, they must conform to the order displayed on the terminal/analytical plane. Viewers have to queue for the terminal plane and in so doing place themselves at an analytical vantage point from which they are able to comparatively count the swings and find the inverse square of time and length. Witnesses have to literally position their heads perpendicularly to the parallel planes of the balls and along the terminal plane that serves as the axis of the demonstrational witnessing field in order to see the balls in a row and to count.

Contingency 4—How to count swings

Let me describe a witnessing account of the pendulum at work if done by a single person. One would first find a way to release three balls with two hands

The Respecification of Galileo's Pendulum

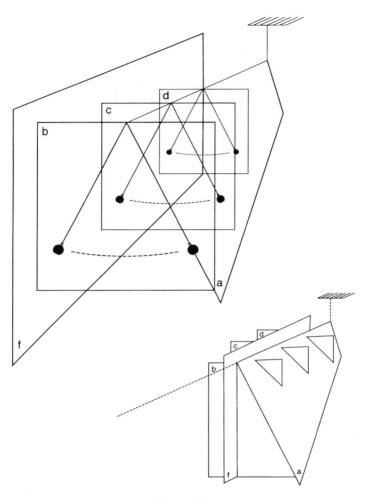

Figure 38

from the same inclined plane *(a)* position, releasing them at the same time and counting (by stepping out of the operational field into the witnessing standpoint after the balls were released) the swings of the longest string. Since it is difficult to determine the full swing by looking at the starting and turning point, the terminal plane is the point of completion of the swing and the standard for counting the swing.[21] After counting two full swings, the observational focus shifts to the entire pendulum, waiting to see if all three balls will meet at the same vertical terminal. The same order applies to the middle and the shortest. The end of observation has to produce the following table of the inverse ratio between the three pendulas' swings and length:

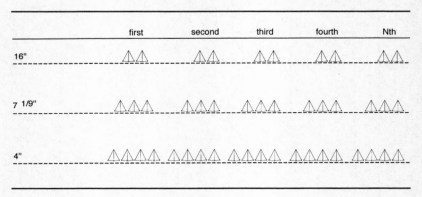

Table 1. Marking Swings in Cohorts of Temporal Synchronies

In the case of a single person observation, three separate observations must be done in order to verify the inverse square rule. I selected one of the three pendula at a time and counted its numbers of swings from the time of the release to the first moment of synchrony with the other two. If the numbers of swings correspond to the ratio 2 : 3 : 4 as 16" : 7 1/9" : 4, it verifies Galileo's inverse square rule.

The successful release of the balls allows them simultaneously to leave the inclined position at an angle that is determined by the demonstrator. The balls swing in different arcs. Though they leave the starting position at the same time, they return in a different time sequence. In this order, the pendulum builds its own time. The temporal placement of each of the balls is situated in the whole temporal pattern of the pendulum. For example, the ball with the longest swinging arc always arrives at the terminal position after the middle and the shortest in the first two swings. The ball with the medium length string arrives after the shortest and before the longest, while the shortest always arrives first, and so on.

Every ball forms its own time row arithmetically comparable to the others. Each swing is a part of its swinging-time row, whereby the latter signifies the account of each of its single constituent swings as a first, third, seventh, nth. Once the time series is established, a comparative measurement can be done in order to determine the ratio between the three swinging-time rows. But counting the swing is the pendulum-organizing work upon whose achievement the swing's respective time-rows are formed, location in the rows found, and the row's comparison provided. Monitoring the release of the ball and counting "first" allows the logical space for the "second" place, which becomes the "second" only because the "first" was already established in memory and the "third" is logically anticipated. The series of logical placements is formed by the observation of the single ball. It is by observing the swing of the single ball that the swing-structured logical location of the ball in the instrument-occupying work is available to itself as the ball's observable time-row. The field of the pendulum's swings that occupies the viewer in a structured way. It is only after the intersubjective structure of this field is established (the oscilloscopic division of labor) that there is any proof of the "order" of motion.

If one releases the balls, lets them swing, and observes the pattern, it becomes clear that there are more and less successful swings, whereby "successfulness" is assessed by reference to seeing the mathematical law of ratio in the demonstration. The best I could achieve was to get synchronic movements only briefly, displayed only for (at best) half a minute, after which the balls acted erratically. This means that at that point, the skills of producing the synchrony disincarnated from the field, leaving the impression that the "demon" of order itself fled the pendulum with the "law" of the thing in its jaw. This suggests that the Gestalt of the ratio is maintained only to the extent that the demonstrator masters the field of the instrument, thereby holding the "demon" of order hostage to it. Therefore, the ratio as an observable *Gestalt* shows that the law of physics is not for all times and for all places, but only for the time set by the skills' unique relation to the instrument. Only in the successful swings is the "ratio" more, or in less-successful swings less visible as a mathematical rule. There are short swings and long swings, fast ones and slow. Every swing is an account of the demonstrator's manipulation of the balls. These demonstrational skills are exhibited as the ball's swing and counted as such. There are many "bad" releases and only a few "correct" ones. "Bad" swings instantly explicate the lack of demonstrational skills and make, like Heidegger's "broken hammer," the demonstrational organization transparent.[22] In contrast, the "right" ones, by displaying the Gestalt of the synchrony of swings, conceal the demonstrational skills and the practical organization of the field. It is the smooth synchronic swing that "transubstantiates" the naive perception into seeing the self-ordered time cohort of the swing. Counting the ball's swings builds the time cohort of the specific ball and makes it analytically comparable with others as a ratio pattern. Seeing the swinging rhythm and not three individual swings is like stroking the strings of a harp in a given order and hearing it as the harmony of the scale, as one. Therefore, it is no longer this or that ball but their observable relations. These relations are not about the balls as things, but about the order of the ball-organizing observations.

Therefore, it can be argued that with "this" equipment Galileo is organizing the practices of making the smooth synchronization of the three balls as a way to demonstrate the law of motion in hanging bodies. In doing so, one can analyze how the swing's work "transubstantiates" naive observation of the instrument's use into seeing the synchronized organization of the swinging balls as a mathematical ratio.

ANALYSIS

Galileo's pendulum organizes the work of time measurement "as the endogenously achieved coherence of pulsed details" which consists of synchronically observing and counting the pendulum's swings.[23] Just as the metronome organizes the count of the tempo, so does the pendulum organize the count of a swing. Swings and counts are paired objects forming a locally embodied workplace through instrumental and bodily synchronization. The embodied production of the swings, hearable as the counting of a swing, is a coherent account of the motion because it is a synchronic pulse. In counting loudly, a coherence is found as "the organizational thing," real, actual, and . . . unremarkably ordinary.[24]

Based upon the above findings, the second part of the Lebenswelt pair of Galilean physics consists of the above-described practical order of the instrument use. Within this use, Galilean physics is performed (in this case, the synchrony of motion as the inverse square root of the strings' lengths) by developing the skillful management of these very instruments. Specifically, the second part of the Livingston pair consists of:

A) The logic of the internal connections between the pendulum (three hanging balls with prespecified inverse ratio lengths), manipulative skills (how to hold three balls with two hands, how to release them simultaneously), and the demonstrated phenomenon of the inverse square ratio.

B) The logic of the internal connections between these instruments, skills, and achieved demonstrational accounts, requires a "tacit" inquiry into the logic itself as a necessary entry into the field of demonstrational objects. Demonstrational instruments, we learn from Galileo, are not "blank" objects but rather, assigned ones; they are found in the order of the demonstrational practice as "instrument-organizing practices." The demonstrational order of the instrument-organizing practices has its own "tacit" logic in need of an inquiry. The inquiry into the inquireable logic of the instrument-organizing practices, the necessary entry into the field of doing physical discoveries, must be successful in order for the natural scientific discovery to occur. This inquiry into the practical logic of Galilean physics is at once a prerequisite for Galilean physics to occur and necessary to the discovery of the second part of Livingston's pair, the structure in which the formal logic of Galilean physics is discovered.

C) The second part of Livingston's pair of Galileo's physics makes the order of the world equipmentally available for mathematical analysis. The unaccounted-for inquiry into the practical logic of the instrument-organizing practices is the discovering structure of Galileo's theorem. Galileo invented a kind of "transubstantiating machine" from this "tacit" inquiry whose "subversive" operation presumes to transform public trust/ordinary witnessing into a mathematical account of the demonstrational field. By giving oneself as a witness to a successful inquiry into the equipment-organizing work of the physicist, this unaccounted achievement is equipmentally "transubstantiated" into the empirical facticity of a natural law—a rationally demonstrable miracle of Renaissance physics. Having achieved proof through the workings of the instrument, Galilean physics allows the possibility for the scientific instrument to reverse the signification of its formal meaning. Instead of signifying only "the proof" in physics it can also be an "object-explicator," an object that explicates its own use as the forgotten constitutive part of the pair.

The Body as *Praxiom* of Galileo's Pendulum

In locating pendular balls, measuring the lengths of the pendular strings, tying a knot, learning how to hold three balls at a different height in the same plane, release them simultaneously, observe swing, count and analyze them, I learned not only the properties of Galileo's pendulum, but also of my own body. I do not mean *body* as pure discursive sign woven into the narrative about the other, nor as time, nor as a phenomenological center of sense experience, but rather as an array of practical contingencies, rules, descriptions, achievements, and failures, all of which have a center in no other time but "now," coherently sustained by the rules of a practical project at hand, and not by any transcendental authority. This praxiological genealogy in a sense recovers the Galileo/Jesuit body.

Neither as sign nor experience, but through learning about my own body in relation to the material and textual demands imposed on it, the body recovers in the realm of pure discourse the conditions for linking the body to the pendulum. In other words, when I ask how to hold three balls with only two hands, how to count three different swings simultaneously, where to look in order to count, how to tie a knot, the Galileo/Jesuit body emerges for me "now" and "here" as nothing else but as my body, yet also as a priori the problem of Galileo's pendulum. I am not suggesting that when involved with the pendulum I am experiencing the same things as a historical subject did, but rather, without falling into the trap of phenomenological reconstruction, I am following the rules of the link, the unavoidable a priori of any pendular operation. Surely, a history of rationality and the motivation to experiment with a pendulum precedes the body-instrument link. However, it does not specify any of the rules, proving only that the body-instrument link retains some degree of autonomy in relation to the pure discourse of rationality.

Thus it is proper to move our discussion of Galileo's pendulum from Foucault (and his discourse on the relation of "body and pleasure") to rules, to learning and to instructive practices. When Koyré described Father Riccioli's experiment, he mentioned it only as an interesting anecdote in the history of measurement and precision. He acknowledged, however, with immense appreciation, the Jesuits' commitment to experimentation for the sake of advancing precision in measurement, and the physical difficulties they encountered. Some of the most dedicated followers of Father Riccioli became too tired to repeat the experiment, and eventually only a few remained in the quest. What held those who stayed together . . . why was their "desire" stronger than the others? Certainly, there are abundant opportunities to psychologize as well as to ignore this question. However, we can answer it properly if we rely on the rules of linking body to instrument as the underlying "motive."

When the body encounters this link, these rules emerge in the course of its inquiry into such linkage as a prediscursive rationality, instructive for the body's transmission of these rules to another body: "*Here* is where to stand!" "*This* is how to tie," "Swing like *this*." A single body could not perform a simple pendular demonstration without inviting the other bodies or dividing itself in time. When the demonstrational field constructs the intelligibility of the site, with all of its internal intricacies, the engaged bodies bond organically around the project as a rational requirement for successful science. Emile Durkheim would call these contingencies "organic solidarity," Foucault "social discipline," but in any event, I know now what such a regime means to my body. And perhaps these contingencies, "organic solidarity" emerging as the intricate rationality of experimental settings, created in a praxiologic sense the conditions for homosocial teams of sexually deprived men to bond analytically and pleasurably during these dry and repetitious operations, otherwise comparable only to their monastic cells. Experimenting, from this vantage point, appears to be a new and momentous mode of bonding. Measuring mechanical time also created the mechanical time of the collectivized experimental procedure; the time to measure was likewise the time of the measurement. In this constructed situation, instrumental restraints added to the need to increase synchrony and induce a higher degree of proximity among the monks' sexually deprived bodies, leading to the excitement of discovery, precision, law, or theory. By re-envisioning Galileo's pendulum in and through these practical details, we have dissolved the discourse of scientific rationality and its implicit "sex" into its formative pedagogy, the *knowing*-body as the living *praxiom*.

Conclusion

In *Civilization and Its Discontents* Freud asserts that "civilization is a process in the service of Eros."[1] Elias, moreover, insists that every civilization must incorporate painful discipline; and for Foucault this painful discipline, by imposing restrictions on our pleasures, is still fundamentally responsible for the invention, intensification and proliferation of these pleasures. If we agree that regulated bodily conduct, not sublimation, is in "the service of Eros" then this book concerns the contribution of Galileo's pendulum, an important artifact of civilizational discipline, to erotic service.

We asked, "How is knowledge related to Galileo's pendulum produced in such a way that the advancement of rationality in Western civilization served the proliferation of new forms of pleasures through rational domination?" We claimed that *mathesis universalis,* adopted by Neoplatonists and applied to the world of sensual experience, became the rational scheme for abstract pleasures of analysis and as such, the impetus of the history of rationality, more or less since Galileo. Unified, explained, classified within grids of mathematical axioms, natural phenomena such as motion, gravity or, light, no longer observed and known by the body, became items in that mathematical catalogue of nature in which all things and relations became subject to the "pleasure of analysis" and domination. In spite of the moral rhetoric of seventeenth-century science, perversion as a ruptured normality colors, to a certain extent, the new ontology of things. The discursive organization of this catalogue changed the sensual nature of natural phenomena. The *place* in the catalogue, rather than a natural *site* shared with the body and senses, determined the *essence* of a thing. This new ontology, which made things what they are by virtue of their discursive location, had detrimental consequences, namely the degradation of the body, which prior to Galileo was the standard of normality and truth. By eliminating a phenomenal *site* and assigning every phenomenon a discursive *place* in this catalogue, mathematics also eliminated the body as a source of order and knowledge, permitting the mathematical catalogue of nature to transcend its prior boundaries and creating a fertile propinquity among all things, even among those previously regarded as naturally irreconcilable.

Ultimately, with the discursive unification of all things in the catalogue, mathematics (like money in a bourgeois society which perverts the natural incongruities and exchanges "hate" for "love," and "stupidity" for "wisdom,") became an abstract unifier, constructing and unifying under the sign of "objectivity" Galileo's law of free fall and Sade's sexual pleasures.

Framing nature in a mathematical catalogue and subjecting its items to the endless "pleasures of analysis," we claimed, seventeenth-century science turned scientific rationality into a "gym" of homosocial exercise, and raised questions about the nature of representational unity under the new mathematical canopy. To the extent that Cartesian exclusion of the body and the feminine from the production of scientific rationality created, particularly in the writings of Descartes, a homosocial epistemology and permitted mathematics to unify a system of representation, we retain merely a partial unity of representation. Once the center of knowledge and pleasure, aesthetics and ethics, the body in the seventeenth century represented a myriad of morally and epistemologically corrupted items, such as "bodily pleasures," "women," "fallen nature," "artisanship," etc., excluded from this unity.

The body, however, was not unintelligible. It became a depository of disqualified knowledges, pleasures and identities, defeated by the formal discourses of rationality and consigned to a condition of muteness. And yet, as the *site* of an experiment or a demonstration, the seventeenth-century experimentalists recognized, as we have seen in the case of Father Riccioli and his faithful Jesuit brothers, the full scope of the body and its role in the production of scientific knowledge. However, this occurred only *before* their work assumed textual form. When experiential physics becomes mathematical text, the body vanishes from the horizons of knowledge. In this regard, since Galileo, mathematical physics has been rendered as a pairing object. One part of the pair consists of the formal and textual structure of physics, while the other part consists of the body-work through which physics and its phenomena come to life. Whenever today we open a textbook of physics, we encounter only half of this science; the other one is mute in the body of the physicist and in its relation to significant instruments, settings, and other bodies. And yet, the mute part of the pair always plays a significant role not only in making scientific discoveries but also in discovering new political technologies.

When Father Riccioli organized his team of experimentalists, he assembled a rare collection of bodies empowered by Christian discourse on sexuality and by sexual deprivation. Subjugated to the chanted counting as a "second code" of language, an arduous task indeed and yet gratifying to the brothers deeply committed to each other, they performed the experiment and gave to humanity a second of time exacted by homoerotic improvisation. "In its mystical or ascetic form," Foucault would have interpreted Riccioli's experiement as a bodily exercise through which the Jesuits ordered, "earthly time for the conquest of salvation."[2] More than any other scientific group of the seventeenth century, the Jesuits discovered how the limitless possibilities of subjection to rationality, their bodies and duration, could be turned into a powerful political technology. Although Riccioli's experiment with Galileo's pendulum may have been a breakthrough not only in mechanics but also in the technology of power, for him and his "loving partners," Irigaray would argue, this would be not only a moment of "fecundity of birth and regeneration, but also the production of a new age of thought, art, poetry, and language: the creation of new *poetics*."[3]

To do justice to the body in the second part of this book we allowed the *knowing* body to represent its relation to Galileo's pendulum; that is to know the pendulum in the bodily mode in which it was known by Father Riccioli and his followers. I'll turn to this topic later. The first part of the book, however, explored the history of homosocial epistemology. We claimed that this history is rooted in the process of desexualization of the body's scientists *via* Neoplatonic discourse on pleasure and, in the case of the Jesuits, Christian asceticism. Not arguing for sublimation, but at the same time not discarding Barthes's penetrating observation that in Western societies "sex is everywhere but in sex" (that is outside the body and in its artifacts), we argued that those technologies, reflected in civilizing artifacts and manuals designed to institute new and reinscribe existing techniques of sexual deprivation, we addressed primarily technologies of time, developed from the erotic performances of those who would subject their sexuality to the operational codes of these crucial technologies. In this respect, instead of sublimation we have prohibited homoerotic pleasures re-emerging, intensified by the very technologies that were supposed to prohibit these desires; resexualizing new mechanics as masculine rationality, they offered a Galilean formula of pleasure: $P = (f) S$, or pleasure is a function of speed. Particularly helpful in unpacking the discourse on sexuality in the history of seventeenth-century scientific rationality were the histories of erotic literature and pornography. They questioned whether or not Neoplatonic sexual deprivation was just another mode of the politics of sexual shame governing, in the guise of epistemology, representational unification by mathematics. While such historians of science as Koyré and Kuhn continued to reproduce this politics of shame by withholding the language of Neoplatonic science from the body and its pleasures, erotic literature and pornography reminded us of the suppressed genealogy of this language, which came, as Rocco put it succinctly, "up the ass."

It should be clear by now to the reader that the purpose of this book was not to write a history of the pendulum as a stable narrative but rather as the tension between the history of sexual shame and a method of desublimation. In so doing I relied on an array of feminist histories of gender and science. The concept of the "handmaid" I found in particular very helpful; it allowed me to argue that the division of scientific labor follows gender divisions established by the homosocial epistemology of seventeenth-century science, on nature and women as one pole, reason and homosociality as the other, regulated by a discourse on sexuality. Although deeply intertwined, sex and gender harbor important discursive domains and distinctions of analysis. While gender studies accept gender as a structural division, they preserve the ontological and substantive status of gender which, in the words of Butler, as "ontological locales" are "fundamentally uninhabitable."[4]

When we adopted the above binaries we agreed that they reflect discursive rules of division rather than ontological ones, which in our case may be used only in the context of the fatal strategy of "repressive desublimation" and the dismantling of the very conditions of sexual divisions. Tethering "women" to "nature," and "ecotheology," essentializes "woman," naturalizing the abjectival circumstances and conditions of the homosocial formation of woman as an identity. And yet the "Death of Nature" does not call for us to resurrect "women," but rather to dismantle "woman" much like homosocial rationality dismantled nature as ontological reality, and by this maneuver dismantle the very conditions of sexual binaries. If we succeed, this may open us to more relevant things, the *work* itself. In order to denaturalize and deconstruct any substantive and ontological sedimentation of gender, an analysis of "sex" focuses on the pleasure of rationality precisely because it is a formative condition of a discourse, which invents gender divisions. Applied to the history of a specific scientific artifact or in general to a history of modern science, pleasure studies, for the reasons outlined above, may also destabilize this history by turning its focus on itself as a pleasure of self-domination. If the history of science is a history of discursive formation and if pleasure is the condition for these formations, then the history of science must be analytically subordinate to pleasure as its formative force. Precisely because pleasure brings substance into life by being naturalized through the subjugation to gender divisions, pleasure as an abstract condition of a discourse of rationality and as an abstract sovereign of all divisions must be constantly open to itself in and as a discourse in order to destabilize the very conditions of any sexual divisions. However, we should not give the wrong impression that we wish the end of politics. Saying "yes" to the deconstruction of a gender identity is not saying "no" to politics, but rather, as Butler convincingly argues, "it establishes as political the very terms through which identity is articulated."[5]

As a concluding note let me briefly refer to the collaboration of Foucault's genealogy and ethnomethodology, a topic that deserves a book on its own. The move from the history of Galileo's pendulum to the *knowing* body as an a priori of this history, and from exposing, as Foucault writes, "a body totally imprinted by history and the process of history's destruction of the body,"[6] to discovering Galileo's pendulum as a uniquely adequate method of the body knowing the pendulum, may be appreciated as a small contribution to the critique of sagging humanism and its "desiring subjects" erected upon the suppressed and homogenized body. Foucault's notion of discourse has been central to our analysis. Foucault presents discourse "as a multiplicity of discursive elements that can come into play in various strategies."[7] It can be both an instrument, he emphasizes, and an effect of power, domination and resistance. It may transmit power but does not always do so, and thus discourse is only the instrumental a priori of a history, knowledge, or power. While in *The Order of Things* and *Archeology* Foucault acknowledges the pleasure of the power of subjugating things to words, in his later work he realized that pleasure of analysis has an oppressive history and that this history leads pleasure back to our relation to the body and pleasure and the civilizational process of the destruction of the body. This change in his perspective occurred in part through his exploration of his own body and pleasures. As a part of the San Francisco gay scene, Foucault got in touch with the practitioners of pleasure, the settings, instruments, skills and communal relations, all of which revealed to him prediscursive formations of pleasures embedded in instructive rather than discursive reasoning.[8] In his late interviews with various gay magazines, Foucault constantly referred to body and pleasures in terms of specific instructive and instrumental actions rather than to the pleasures governing pure discourse. There, Foucault begins to resemble "Agnes." Like "Agnes," Foucault eroticized his own body in order to *know* pleasures as body and not pure discourse, confronting here *via* pleasures a different kind of a priori. It is only at this juncture of the two eroticized bodies with a pedagogy of pleasures, that I could conceive a methodological relation between Foucault and "Agnes," that is, between genealogy and ethnomethodology.

Like "Agnes" and Foucault I used my body to know Galileo's pendulum in its practical rather than discursive specificity. While Galileo's theory about the pendulum and his textual instructions about the specific pendulum came to me as a historic and textual artifact, they only promised that the pendulum would work in accordance with Galileo's claim. Finding the necessary material, making the pendulum, and demonstrating its laws, belongs to the present and to my own bodily skills. In the process of making the pendulum I encountered numerous rules regulating the relation between the body and the pendular materials, not only to assemble the instrument but also to make it work as promised in the text. At all points in this arduous enterprise, the body discovered the entire and unaccounted for "physics" of the link in making and using the pendulum. Gravity, as the main force exhibited in the isochronic motion of the pendulum, can come into visual prominence only after the body-instrument link has been established, and not in any random fashion, but only in a ruled, governed way, and accessible only to the body in inquiry. And yet, how to hold three hanging balls at different lengths with only two hands hides a rule, which cannot be discursively defined, only known through patient performance. However, this is not to say that nothing can be said about this or any other rule regulating the body-instrument link. Only that discursive definition will not do. "Hold like *this!*" or, "Release the balls from *this* point!" "Stand *there* and look *here!*" Instructive utterances of this kind usually help more. The language of discourse must transform into the language of instructions in order to link the body to the instrument and deliver what Galileo's text only fantasizes about. This is the difference between pleasure of pure discourse and the pedagogy of the *knowing* body.

When I struggled with a piece of string to tie a knot right on the point where mathematically should be the fixed point of a pendulum, the practical, bodily and transcendental aspects of Galileo's pendulum intersected with geometry just as the gaze of a women intersected a divine ray of light coming from the window at the fixed point of a balanced scale measuring gold in Jan Vermeer's famous *Woman Holding a Balance*. In the astonishing essay "Abriosia and Gold," Michel Serres respecifies Descartes's geometry of a "dead eye" through the gaze of the woman in this painting.[9] Although enshrined in bodily pleasures and pains by being pregnant, she peacefully suspends the scale above the gold spread over the table, while behind her head, significantly on display, a depiction of the *Last Judgment* hangs on the wall. Through this arrangement, Vermeer suggests that the body pleasures, salvation, and judgment can balance each other at the fixed point of a scale whereby the point is both an object in mechanics and a sign in ethics. As the geometric center of the painting, the fixed point of the scale, a sort of a pendulum, is "the static pole of equality and balance—the golden mean, the referent-point of Cartesian space." In the space this fixed point signifies is the implicit "presence of the nonreal within the real, a spatial element absent from space."[10] In his painting Vermeer submits the idea that Western civilization is not advanced by the division of pure cogito and bodily pleasures, but by discovering the intersection of the two, or in viewing the two as aspects of a fixed point. Indeed, Serres argues, with the discovery of the fixed point we moved from the object to a sign, enabling the representation of nature to become discursive. But discourse, in turn, becomes representable as well, for as with the pendulum, the counter-swing balances the initial swing.

Like us, Serres affirms that the pendulum provides a model for space, time, and its measurement in both nature and society. By organizing itself around a simple machine such as the pendulum, he insists, society "discovered the practical efficiency of a fixed point, determined once for all, around which a given force multiplies its power."[11] Following Foucault, the balanced scale may be read as one's relation to the two aspects of power in which balancing may have been an important part of our technology of self.

Now I understand why I could never tie the knot in the right place: this point was not in the given space. Trying to *find* the point with my body was like trying to get into the space from which the body was excluded. Were there not politics to this geometry? In my lonely imperfection, I remained within the space dominated by my body. As I was constructing my pendulum as a sign from antispace into the space of a body, I likewise surrounded and fortified myself with the pendulum's wholly embodied geometry. My body became disciplined and observant of the rules in this mathematical space. As I became increasingly disciplined and observant, the performance of the pendulum improved. By incorporating the geometry of the pendulum into my bodily conduct I saw myself as I am seen through the "dead eye," a body positioned among and defined by lines, angles, and planes. Is this a defeat or an ultimate domination of the rational by the body? Perhaps neither, perhaps it is a balance in seeing. I saw with the living eye how am I seen through the "dead eye." And this moment of practice seems to be the *only* point of seeing and of a true closure.

Notes

FOREWORD

1. Bjelić picks up one of Garfinkel's characteristic expressions—"demonic contingencies"—and plays on the word "demon-stration" to signal the way unanticipated, often rather wild and disconcerting, circumstances haunt the effort to reproduce the law.
2. The demonstration was described in unpublished drafts of a manuscript that Garfinkel circulated in 1988 and 1989 (Garfinkel et al. 1989).
3. Eric Livingston (1995) also used Galileo's pendulum demonstration in an ethnomethodological respecification of physical experimentation. Livingston's treatment was developed independently of Bjelić's, and provides an interesting companion piece with a somewhat different lesson.
4. The "problem of replication" (Collins 1985) is well known in science studies circles. Not only does it apply to contemporaries who attempt to replicate the experiments of colleagues and rivals in their specialized fields, it applies with extra force to efforts to replicate experiments across centuries of history. The problem of replication is both a practical and rhetorical one. Practically, it can be difficult to reproduce exactly the materials and practices of scientists who are long dead. Theories and observation languages are likely to have changed, and the original equipment may be difficult and expensive to find or rebuild.
5. This point about the unusually "robust" character of Galileo's demonstration was made by Bruno Latour, in a personal communication during a seminar at the Centre for the Sociology of innovation, at l'Ecole Nationale Superieure des Mines de Paris, May 1998. A group of faculty and students at the seminar collaborated in an effort to perform the inclined plane demonstration as described by Galileo and Bjelić.
6. Garfinkel (2002) uses the inclined plane demonstration to demonstrate the possibility of "losing the phenomenon"—a practical possibility that he claims is essential to the natural, but not the social, sciences. Less no robust; and more technically exacting and historically questionable, demonstrations tend to make a different possibility perspicuous. We might call the latter "not getting to first base." Far from losing the phenomenon, one is left in doubt as to whether it is possible to "get" it in the first place. I gained insight about the difficulties associated with *getting the phenomenon* by participating in an International Workshop on Replications of Historical Experi-

ments in Physics, Their Function in History, Philosophy, and Sociology of Science and in Science Teaching, Carl von Ossietzky University of Oldenburg, Oldenburg, Germany (24–28 August 1992).

INTRODUCTION

1. Julia Kristeva, *Desire in Language: A Semiotic Approach to Literature and Art,* ed. Leon S. Roudiez (New York: Columbia University Press, 1980), 165.
2. Donna Haraway, *Primate Visions: Gender, Race, and Nature in the World of Modern Science* (New York and London: Routledge, 1989), 2.
3. Peter Singer, review of *Dearest Pet: On Bestiality* by Midas Dekkers. http://www.nerve.com/opinions/singer/heavypetting/main.asp.
4. Ibid.
5. For more on interspecies sexuality see Piers Beirne, "Towards a Concept of Interspecies Sexual Assault," *Theoretical Criminology* 1, no. 3 (August 1997): 318.
6. Haraway, *Primate Visions,* 2.
7. The learned volumes, written and read; the consultations and examinations; the anguish of answering questions and the delights of having one's words interpreted; all the stories told to oneself and to others, so much curiosity, so many confidences offered in the face of scandal, sustained—but not without trembling a little—by the obligation of truth; the profusion of secret fantasies and the dearly paid right to whisper them to whoever is able to hear them; in short, the formidable "pleasure of analysis" (in the widest sense of the latter term) which the West has cleverly been fostering for several centuries: all this constitutes something like the errant fragments of an erotic art that is secretly transmitted by confession and the science of sex.

Michel Foucault, *The History of Sexuality: An Introduction,* vol. 1, trans. Robert Hurley (New York: Vintage Books, 1990), 71.

8. Judith Butler explains the Nietzschean notion of creative value: "Value emerges as the 'show' of strength or superior force and also comes to conceal the force relations that constitute it; hence, value is constituted through the success of strategy and domination; it is also that which tends to conceal the genesis of its constitution." Judith Butler, *Subjects of Desire: Hegelian Reflections in Twentieth-Century France* (New York: Columbia University Press, 1999), 181.
9. Butler, *Subjects,* 229.
10. Friedrich Nietzsche, *Thus Spoke Zarathustra: A Book for None and All,* trans. Walter Kaufmann (New York.: Penguin Books, 1978), 115.
11. Butler, *Subjects,* 231. As Nietzsche poetically marks the simultaneous birth of the self and rationality: "The self says to the ego, 'Feel pleasure here.' Then the ego is pleased and thinks how it might often be pleased again—and that is why it is *made* to think." Nietzsche, *Thus Spoke Zarathustra,* 35.
12. Eve Kosofsky Sedgwick, *Between Men, English Literature, and Male Homosocial Desire* (New York: Columbia University Press, 1985), 1–5; *Epistemology of the Closet* (Berkeley: University of California Press, 1990), 87–88.
13. Butler, *Subjects,* 182.

14. Michel Foucault, *The Use of Pleasure: The History of Sexuality,* vol. 2, trans. Robert Hurley (New York: Vintage Books, 1990), 151.

15. Terry Eagleton, *Ideology of the Aesthetic* (Cambridge: Basil Blackwell, 1990), 394.

16. Mario Biagioli, *Galileo Courtier: The Practice of Science in the Culture of Absolutism* (Chicago and London: The University of Chicago Press, 1993).

17. "Galilean science" is used here as a broader term then "Galileo's science." While the latter stands for a historical model the former stands for an ideal type, a pure logical model of positive science as formulated and criticized by Edmund Husserl in *The Crisis of European Sciences and Transcendental Phenomenology* (Evanston, Ill.: Northwestern University Press, 1970).

18. Ian Hunter, "Aesthetics and Cultural Studies," in *Cultural Studies,* ed. L. Grossberg et al. (New York and London: Routledge, 1992), 358.

19. The scientist may object that the pendulum is not a scientific instrument but rather a simple object. While agreeing with this objection I use "instrument" not to emphasize its relation to *episteme* but rather to the body and discipline. Like a baton, it is a simple object, yet it is regarded as an instrument of order.

20. I am thankful to Branka Arsić for pointing out to me Descartes's example of the "dead eye" and its relevance to his sexualization of perception. See René Descartes, *Discourse on Method, Optics, Geometry, and Meteorology,* trans. Paul J. Olscamp (New York: Bobbs-Merrill Company, Inc., 1965), 91.

21. Ibid.

22. Jacques Lacan, *The Four Fundamental Concepts of Psychoanalysis,* ed. Jacques-Alain Miller, trans. Alan Sheridan (New York and London: W.W. Norton & Company, 1998), 224.

23. Ibid.

24. My discussion of Descartes's sexualization of perception closely follows Branka Arsić's exposition of Descartes in *Razum i Ludilo: Neki Aspekti Dekartovih Meditacija o Prvoj Filozfiji* (Reason and Madness: Some Aspects of Descartes's Meditations on First Philosophy) (Belgrad: Stubovi Kulture, 1997), 135–83.

25. Descartes, "Letter to Vatier from 22 Feb 1638," in *The Philosophical Writings of Descartes,* vol. 2 of *Correspondence,* trans. John Cottingham, Trobert Stoothoff, Diugald Murdoch and Anthony Kenny (Cambridge: Cambridge University Press, 1984), 86.

26. Arsić, *Razum i Ludilo,* 182.

27. Mario Biagioli, "Knowledge," *Freedom and Brotherly Love: Homosociality and the Accademia dei Lincei, Configurations,* 1995, no. 2, 141.

28. Desiree Hellegers, *Handmaid to Divinity: Natural Philosophy, Poetry, and Gender in Seventeenth-Century England* (Norman: University of Oklahoma Press, 2000).

29. As it will become clear "the body" here is always a shorthand for the set of describable situated practices and prediscursive knowledge, or the "body of practices" rather than a natural and predetermined center of thinking or acting. This point is further clarified in the second part of the book.

30. *Technologies of the Self: A Seminar with Michel Foucault,* ed. Luther H. Martin, Huck Gutman, and Patrick H. Hutton (Amherst: The University of Massachussetts Press, 1988).

31. Arnold Davidson, *The Emergence of Sexuality: Historical Epistemology and the Formation of Concepts* (Cambridge, MA: Harvard University Press, 2001), 212.

32. Sigmund Freud, *New Introductory Lectures on Psychoanalysis* (New York: W. W. Norton, 1933), 133.

33. Virginia Woolf, *A Room of One's Own* (San Diego, New York, and London: Harcourt, Inc., 1981), 103.

34. "Reconstructivist" is a loose term for that cadre of historians who adhered to the method of the reconstruction of Galileo's experiments as a means of interpreting Galileo's work. This method was developed as an a posteriori test for the hypothesis formulated by Alexandre Koyré about Galileo's work. The most prominent role in this group belongs to the work of Stillman Drake, *Galileo at Work* (Chicago: University of Chicago Press, 1981); "Galileo's Discovery of the Law of Free Fall" *Scientific American* 228, no. 5 (May 1973): 84–92; "The Role of Music in Galileo's Experiments," *Scientific American* 232 (June 1975): 98–104; *Galileo Studies: Personality, Tradition, and Revolution* (Ann Arbor: The University of Michigan Press, 1970); *Galileo: Pioneer Scientist* (Toronto: University of Toronto Press, 1990); "Mathematics and Discovery in Galileo's Physics," *Historica Mathematica* 1 (1974): 129–50; "Analysis of Galileo's Experimental Data" *Annals of Science* 39 (1982): 389–97; "New Light on a Galilean Claim About Pendulums" *Isis* 66 (1975): 92–95; "Velocity and Eudoxian Proportion Theory" *Physics* 15 (1973): 49–64; Stillman Drake and James MacLachlan "Galileo's Discovery of the Parabolic Trajectory" *Scientific American* 232 (March 1975): 102–10.

35. Koyré writes: "A bronze ball rolling in a 'smooth and polished' wooden groove! A vessel of water with a small hole through which it runs out and which one collects in a small glass in order to weigh it afterwards and thus measure the times of descent (the Roman water clock, that of Ctesebius, had already been a much better instrument): what an accumulation of sources of error and inexactitude!

It is obvious that the Galilean experiments are completely worthless: the very perfection of their results is a rigorous proof of their incorrection." "An Experiment in Measurement," *Proceedings of the American Philosophical Society* 97, no. 2 (April, 1953): 224.

36. Robert Markley, *Fallen Languages: Crisis of Representation in Newtonian England, 1660–1740* (Ithaca and London: Cornell University Press, 1993), 232. For an extensive account of Hook's role of a laboratory technician, see an excellent account by Steven Shapin, *A Social History of Truth: Civility and Science in Seventeenth-Century England* (Chicago: Chicago University Press, 1994), 378–89.

37. Hellegers, *Handmaid to Divinity,* 131.

38. Bruno Latour, *Pandora's Hope: Essay on the Reality of Science Studies* (Cambridge: Harvard University Press, 1999), 215.

39. Harold Garfinkel, *Studies in Ethnomethodology* (Englewood Cliffs, N. J.: Prentice-Hall, 1967).

40. Sigmund Freud, *On Dreams,* trans. and ed. James Strachey (New York and London: W. W. Norton & Company, 1989), 44.

41. Harold Garfinkel, Eric Livingston, Michael Lynch, and Doug MacBeth, "Respecifying the natural sciences as discovering sciences of practical action, I and II: Doing so ethnographically by administering a schedule of contingencies in discussion

with laboratory scientists and by hanging around their laboratories" (unpublished manuscript, 1988). See also Michael Lynch, *Art and Artifact in Laboratory Science* (London: Routledge & Kegan Paul, 1987), and Eric Livingston, *The Ethnomethodological Foundation of Mathematics* (London: Routledge & Kegan Paul, 1986).

42. Foucault, *The History*, 61.

43. Peter Dear, *Discipline and Experience: The Mathematical Way in the Scientific Revolution* (Chicago: The University of Chicago Press, 1995); also, "From Truth to Disinterestedness in the Seventeenth Century," *Social Studies of Science* 22 (1992): 619–31.

44. Michel Foucault, *Discipline and Punish: The Birth of the Prison*, trans. Alan Sheridan (New York: Vintage Books, 1979), 137.

CHAPTER 1. TIME, PLEASURE, AND KNOWLEDGE

1. Gilles Deleuze, *Masochism: An Interpretation of Coldness and Cruelty*, trans. Jean McNeil and Aude Willm (New York: Zone Books, 1991), 117.

2. Thomas S. Kuhn, *The Structure of Scientific Revolution* (Chicago: University of Chicago Press, 1970), 120.

3. Guy Debord, *The Society of the Spectacle* (New York: Zone Books, 1994).

4. Stuart Sherman, *Telling Time: Clocks, Diaries, and English Diurnal Form, 1660–1785* (Chicago: University of Chicago Press, 1996), 2.

5. Norbert Elias, *Time: An Essay*, trans. (from the German) Edmund Jephcott (Oxford and Cambridge: Blackwell, 1992), 114.

6. Elias, *Time*, 45.

7. Who other than Marx can offer clearer insight into the political economy of time? Slicing human labor in equal temporal standards has merged the mechanics of time and the body into a synchronic *dance* of capitalism. Being the measure of all labor, mechanical time has, as Marx insists, radically changed the logic of value, forcing the economist to accept "the economy of time" as the "first economic law." He writes:

> In the final analysis, all forms of economics can be reduced to an economics of time. Likewise, society must divide up its time purposefully in order to achieve a production suited to its general needs; just as the individual has to divide his time in order to acquire, in suitable proportions, the knowledge he needs or to fulfill the various requirements of his activity. (Karl Marx, *Grundrisse*, in *Selected Writings*, ed. David McLellan [Oxford: Oxford University Press,1977], 362)

By tethering time to labor, mechanical time became a standard measure of an exchange between two or more concrete labors done in "creative time," and it was decomposed and recomposed as an abstract time by this exchange and subsequently measured by money. See Marx, "On Money," in *Selected Writings*, 109. In the same way Galileo's pendulum subordinates phenomenal space, concrete objects, and their motions to the abstract concept of time, so does Marx's theory of *general equivalence* standardize labor by means of mechanical time, and subordinate the concrete existence of a society to the abstract rules of general exchange and their logic of value. Eric Alliez concludes that in the capitalist mode of production, therefore,

[E]verything is reversed, for abstract, homogeneous time, the measure of the exploitation and subsumption of the *socius* under the regime of equivalence (time is this regime's very matter), is undoubtedly opposed to every idea of a creative duration, though it invokes creative duration as its natural complement (be it only for this: *the subsumption of society has turned itself into the production of society*), just as a science calls on the metaphysics that founds it. (Eric Alliez, *Capital Times: Tales from the Conquest of Time*, trans. Georges Van Den Abbeele [Minneapolis: University of Minnesota Press, 1996], p. xviii)

The universal standardization of labor by virtue of mechanical time led to the formation of universal standards of exchange which transformed, in Marx's view, the natural aggregate of body and labor into the universe of abstract laws of political economy. Like Galileo's physics, political economy measures and abstracts the concrete world into a discourse in order to understand, explain, and regulate it. Like Galileo, Marx argues that political economy, in order to conduct an objective analysis, must deduce particular bodily time out of an abstract time on the basis of which an axiomatic order of the universe of labor may be scientifically constructed. The abstraction of human labor and the formation of political economy as an objective science, followed the same economy of "sex."

As in Galileo's mathematical physics, political economy forms its concepts and conducts its inquiry by polarizing rational discourse and nature along sexual lines. No different from the Greek sexualized metaphysic of *form* and its masculine domination over the feminine *matter,* such representational formation rests upon oppositional agencies. It is not surprising, therefore, that Jean-Joseph Goux finds in Marx's theory of *general equivalence* not only a paradigm of modernity but also a mode of symbolization that incorporates both economic and sexual signifiers. He writes:

In short, I came to affirm that the *Father* becomes the general equivalent of subject, *Language* the general equivalent of signs, and the *Phallus* the general equivalent of objects, in a way that is structurally and genetically homologous to the accession of a unique element (let us say, *Gold,* for the sake of simplicity) to the rank of the general equivalent of products. Thus, what had previously been analyzed separately as phallocentrism (Freud, Lacan), as logocentrism (Derrida), and as the rule of exchange by the monetary medium (Marx), it was now possible to conceive as part of a unified process. Questions such as those of the third, of the Other, of the exclusion of the One measure, found their structuring places in this logic of exchange. (Jean-Joseph Goux, *Symbolic Economies: After Marx and Freud,* trans. Jennifer Curtiss Gage, [Ithaca, N.Y.: Cornell University Press, 1990], 4)

The polarities of the abstract and the natural engendered by the new principles of exchange led to, on the one hand, the unification of the representational technologies otherwise distant in terms of economy and sexuality, and, on the other hand, subjugation of the *material* of life to the sexualized *idea* of life. Mechanical time has invented the social mechanics of this hierarchy.

8. Alliez, *Capital Time,* xix.

9. Edward T. Hall, *The Dance of Life: The Other Dimension of Time* (Garden City, N.Y.: Anchor Press, Doubleday, 1983), 140.
10. Foucault, *Discipline and Punish: The Birth of the Prison,* trans. Alan Sheridan (New York: Vintage Books, 1979), 135.
11. Norbert Elias, *Involvement and Detachment,* trans. Edmund Jephcott (New York: Basil Blackwell, 1987), 4.
12. Elias, *Involvement,* 11.
13. Foucault, *Discipline,* 135.
14. Plato, *Laws,* 1.636d–e; in Foucault, *The Use of Pleasure: The History of Sexuality,* vol. 2, trans. Robert Hurley (New York: Vintage Books, 1990), 57.
15. Foucault, *The Use of Pleasure.*
16. Ibid., 59.
17. Aristotle, *Physics* (New Brunswick: Rutgers University Press, 1998), 119–33; Koyré, *Metaphysics and Measurement: Essays in Scientific Revolution* (Cambridge, Mass.: Harvard University Press, 1968), 1–15; A. C. Crombie, *Augustine to Galileo: The History of Science A.D. 400–1650* (Cambridge: Harvard University Press, 1953), 212–73.
18. Aristotle, *Physics,* 122.
19. Gregory of Nyssa, *De hominis opificio* 22.7, quoted in Peter Brown, "Bodies and Minds: Sexuality and Renunciation in Early Christianity." In *Before Sexuality: The Construction of Erotic Experience in the Ancient Greek World,* ed. David M. Halperin et al. (Princeton: Princeton University Press, 1990), 487.
20. William H. McNeill, *Keeping Together in Time: Dance and Drill in Human History* (Cambridge: Harvard University Press, 1995); Eugene Louis Backman, *Religious Dances in the Christian Church and in Popular Medicine* (London: George Allen & Unwin, 1952), 37.
21. McNeill, *Keeping,* 76.
22. David S. Landes, *Revolution in Time: Clocks and the Making of the Modern World* (Cambridge: Harvard University Press, 1983), 63.
23. Landes, *Revolution,* 69.
24. Louis Mumford, *Technics and Civilization* (New York: Harcourt, Brace & World, 1963), 14–15.
25. "Canonical hours" applied not to points of time but rather to a division of a day along bands of time. Romans called them *tierce* (from sunup to midmorning), *sext* (from then to midday), *none* (from midday to midafternoon), and vespers (from then until nightfall). See Landes, *Revolution,* 404 n. 22.
26. J. D. North, "Monasticism and the First Mechanical Clocks." In *The Study of Time II, Proceedings of the Second Conference of the International Society for the Study of Time,* ed. J. T. Fraser and N. Lawrence (Berlin and New York: Springer-Verlag, 1975), 382.
27. Landes, *Revolution,* 62.
28. Jacques Le Goff, *Time, Work and Culture in the Middle Ages* (Chicago: University of Chicago Press, 1982), 39.
29. Foucault, *The History of Sexuality: An Introduction,* vol. 1, trans. Robert Hurley (New York: Vintage Books, 1990), 63.
30. Landes, *Revolution,* 63.

31. The sleepyheads were probed out of bed and urged to the office: also probed during service if they failed in their obligations. "Where the flesh was weak, temptation lurked." Demons were being blamed for seducing monks to sleep. Raoul Glaber (early eleventh century) tells about the demon who seduced the monk "by holding out the lure of sweet sleep." He articulates the demon: "As for you, I wonder why you so scrupulously jump out of bed as soon as you hear the bell, when you could stay resting even unto the third bell. . . . But know that every year Christ empties hell of sinners and brings them to heaven, so without worry you can give yourself to all the voluptuousness of the flesh." (Landes, *Revolution,* 65)

The inability to achieve control over the bodily pleasures was an issue for the monastic community to address and to moralize, to record and to publicize the self-examination and confession of the weaknesses of the will; inability to conform to the sound of the "bell," to the command of the mechanical time became a historical event worthy of record. For example, to the ordinary monk probably the hardest aspect of monastic discipline was to wake up in the middle of the night and pray. The notion of "reforming" the house meant exactly the imposition of this duty (Landes, 65). Or, on another occasion, a monk woke up thinking that he heard the bell, he ran to the chapel but nobody was there. He returned to the dormitory and saw other monks sleeping soundly, he then understood that "this was all a temptation of the devil, who had awakened him at the wrong time, so that when the bell for nocturnes really rang, he would sleep through it" (65).

32. North, "Monasticism," 384.

33. Alexander Waugh, *Time, Its Origin, Its Enigma, Its History* (New York: Carroll & Graf Publisher, Inc., 2001) 26.

34. F. C. Haber, "The Cathedral Clock and the Cosmological Clock Metaphor," in *The Study of Time II, Proceedings of the second conference of the International Society for the Study of Time,* 404.

35. Ibid., 407.

36. Landes, *Revolution,* 16.

37. Mumford, *Technics,* 14–15.

38. Sherman, *Telling Time,* xi.

39. Benedict Anderson, *Imagined Communities* (London: Verso, 1991), 24.

40. Dr. Iwan Bloch, *Anthropological Study in Strange Sexual Practices in All Races of the World,* trans. Keene Wallis (New York: Falstaff Press Inc., 1933), 39–40.

41. Landes, *Revolution,* 92.

42. Ibid., 269.

43. Allucquere Rosanne Stone, *The War of Desire and Technology at the Close of the Mechanical Age* (Cambridge, Mass.: The MIT Press, 1995), 7.

44. Paul Virilio, *Open Sky,* trans. Julie Rose (London: Verso, 1997), 111.

45. Ibid., 11.

46 Ibid., 104.

47. Ibid.

48. C. G. Ballard, *Crash* (New York: Vintage Books, 1974), 6.

49. Vivilio, *Open Sky,* 116.

CHAPTER 2. THE PERVERSION OF OBJECTIVITY AND THE OBJECTIVITY OF PERVERSION

1. Peter Dear observes that "nowadays, 'objectivity' tends to be conceptualized in terms of its opposite, 'subjectivity.'" See Dear, "Truth to Disinterestedness in the Seventeenth Century," *Social Studies of Science* 22 (1992): 619–31.

2. Lorraine Daston, "Objectivity and the Escape from Perspective," *The Science Studies Reader,* ed. Mario Biagioli (New York: Routledge), 110–23.

3. Robin May Schott, *Cognition and Eros* (Boston: Beacon Press, 1988), x.

4. Steven Shapin, *A Social History of Truth: Civility and Science in Seventeenth-Century England* (Chicago: University of Chicago Press, 1994), 148–54.

5. James Bono, *The Word of God and the Languages of Man: Interpreting Nature in Early Modern Science and Medicine* (Madison: University of Wisconsin Press, 1995).

6. Julie Robin Solomon observes that these recent works, inspired by Michel Foucault, have "begun to unearth the historical 'layers' that make up our modern idea of objectivity." See Solomon, *Objectivity in the Making* (Baltimore: John Hopkins University Press, 1998), 9. She also stresses how these studies demonstrate the moral function of this concept. She writes: "Mechanical and aperspectival objectivity appears as significant 'moral' remedies in the nineteenth century" (9). The view that "objectivity" coextends with morality is based on the view that truth and virtue are connected. Alasdair MacIntyre champions this position. He writes: "Every practice requires a certain kind of relationship between those who participate in it. Now the virtues are those goods by reference to which, whether we like it or not, we define our relationship to those other people with whom we share the kind of purposes and standards which inform practices." See MacIntyre, *After Virtue: A Study in Moral Theory* (Notre Dame, Ind.: University of Notre Dame Press, 1981), 178–79. "Objectivity" would be that virtue that defines relationships among scientists in order to establish firm sets of purposes and standards for scientific practices. But this is a whole lot to assume about science in advance by holding on to the conventional view of science as set practices of normalization; what scientific practices follow as their informed standards, purposes, and truths often pervert moral standards.

7. Margaret C. Jacob, "The Materialist World of Pornography," in *The Invention of Pornography: Obscenity and the Origins of Modernity, 1500–1800,* ed. Lynn Hunt (New York: Zone Books, 1993), 158.

8. Paula Findlen, "Humanism, Politics and Pornography in Renaissance Italy," in *The Invention of Pornography: Obscenity and the Origins of Modernity, 1500–1800,* ed. Lynn Hunt (New York: Zone Books, 1993), 59. See also Edgar Wind, *Pagan Mysteries in the Renaissance* (New York: Barnes & Noble Inc., 1968), 157; Leonard Barkan, *Transuming Passion: Ganymede and the Erotics of Humanism* (Stanford, Calif.: Stanford University Press, 1991), 71–72; Paul Oskar Kristeller, *Studies in Renaissance Thought and Letters* (Roma: Edizionbe di Storia e Letteratura, 1956), 113–14.

9. Lynne Lawner, Introduction to *I Modi: The Sixteen Pleasures: An Erotic Album of the Italian Renaissance,* ed. and trans. Lynne Lawner (Evanston, Ill.: Northwestern University Press, 1988), 19.

10. Jean Baudrillard, *Seduction* (New York: St. Martin's Press, 1990) 29.

11. Jacob, "The Materialist World of Pornography," 158.

12. The relationship between the seventeenth-century mechanics and pornography could be explored beyond the latter being just a satire of the former. Although pornography stresses the value of nature and the senses, and on the surface appears to be a *reversed* antidote to Neoplatonic epistemology, we could argue that seventeenth-century pornography had been invented around "sex" and the mechanics of the Cartesian body, as an explicit object of its representation, while seventeenth-century natural philosophy concealed this significant connection with "sex" as the hidden economy of its representation. The Cartesian dualism of the world of the senses and pleasure, or *res extensa,* and the world constructed by pure reason, or *res cogitans,* was satirically perverted by seventeenth-century pornography. With this "fatal strategy" seventeenth-century pornography employed discourse of pure rationality—*res cogitans*—to invent, expand, and intensify pleasure. In this way seventeenth-century pornography made explicit the hidden economy of the analytic power of scientific rationality and the enormous potential of Cartesian rationality for the proliferation of pleasure. Thus it should also be regarded as a form of cultural criticism of seventeenth-century science.

There are historical and representational parallels here. For example, David Fox argues that pornography emerged in the middle of seventeenth century when sex became intellectualized, experiences recorded, and perversions catalogued. Pornography became "heretically Cartesian, Newtonian or Hobbesian . . ." (in Jacob, "The Materialist World of Pornography," 165). Jacob attests that seventeenth-century pornography represented the human body by the same transcendental logic as the new mechanics did. Much as in the physical universe of the new mechanics of the time, "bodies were stripped of their texture, color and smell, of their qualities, and encapsulated as entities in motion, whose very beginning is defined by that motion" (Jacob, 164). Pornographic space, time, motion and body, Jacob claims, are "abstraction[s] divorced from the everyday space in which are seen only the appearances or qualities of bodies clothed and decorated, disguised by color and texture, bodies visible to the public eye, encoded with the actual in imagined symbols of status, power and sexuality" (Jacob, 182). Pornographic dialogues contained mechanistic expressions such as, "the long engine with the most splendid pleasure in the world" (166), or sexual intercourse between male and female was represented as mediated by an electricity which would move man and women into each other (175). Significantly, "on occasion," as Jacob points out, "pornographers even credited Hobbes or Descartes with having inspired them" (167). The concept of a personal love was replaced by an apersonal passion.

The overlap between the seventeenth-century mechanics and pornography allowed for the Cartesian logic of mechanics to serve as a tool for *coding* sex as a pleasure, that is, for inventing "sex" as a law of mechanics. There are three aspects of this relationship: mechanization of sex, its instrumentation, and its medicalization.

The mechanization of sex not only helped to explain but also to intensify "sex." In the French eighteenth-century masterpiece of pornography *Therese philosophe,* Therese insists on living out the implication of the laws of nature because, as Jacob explains, "The mechanization of nature holds the key to passion, just as passion verifies nature's laws" (Jacob, 181). This seems to explain what pornography intuitively understood and what the new mechanics did not, namely, that the new mechanics is a theory of a rational relation, actions and reactions, which hides or obscures the *code* of intensified and controlled pleasure. Therese explains: "It is the arrangement of organs, the dispositions of the fibers, a certain movement, the liquids, which make up the genre of the passions . . . nature is uniform" (181). Thus, in anal intercourse, Jacob continues, "Therese quickly understands that she must also move, to create "un mouvement opposé." The senses operate mechanically and, from the physical experience of good and evil, humans can derive moral definitions of good and evil (181). It is interesting to notice that Foucault himself takes the mechanistic view of pleasure. For example, when he invokes "fisting" as an alternative technique for inducing polymorphous pleasures against orgasm, he refers to pleasures of what Gayle Rubin characterizes as "seducing one of the jumpiest and tightest muscles in the body." See Gayle Rubin, in *Saint Foucault: Towards a Gay Hagiography* by David M. Halperin (Oxford: Oxford University Press, 1995), 91.

The instrumentation of "sex" elevated the art of making sexual instruments into a new referent of pleasure above "nature." This is evident in the seventeenth-century play about the art of dildo, *Sodom,* by Rochester. Entering a dildo shop, Officina, one of the women who craves more pleasure, addresses the owner and the maker of dildos: "Let's see the great improvement of your art; The simple dildos are not worth a fart" (E. of R. *Sodom: A Play* [Antwerp, 1984], 45). Or, elsewhere in her quest, she articulates the art of instrumentation as the art of increasing pleasure: "Oh fie! Scarce exceed a virgin span / Art should exceed what nature gave to man." In this regard, as Jacob states, "passion is so much the contrivance of artifice" (Jacob, 180). Buggertanthos, in *Sodom,* explicates the role of dildos in the universal exchange of pleasures: "For I had once a passion for a Horse / Now wand'ring o'er this vile cunt starving land I am content with what comes next by hand" (*Sodom,* 40). Elsewhere Pockenello claims: "Now the dictates of our sense pursue / We study pleasures still and find new" (*Sodom,* 30). Similarly, we read in the epilogue a condemnation of the feebleness of nature and potency of arts: "Damn'd feeble pricks, we hate them, they're but toys; We're for the more substantial solid joys / Of a brave stiff romantic swinging Prick / That's twice five inches long and seven thick" (*Sodom,* 57). It is also interesting that while women were attached to and craved dildos, the king preferred boys. This division of sexual pleasures in the time of the birth of mechanics, reflects homoerotic structures of patriarchal society and marginalization, control and intensification of female sexuality through the art of instrumentation.

With respect to the medicalization of "sex" Rochester's sexual imagination about the relation between the instrument and sex found its mechanical codification in the history of the vibrator. Although the vibrator was invented in the nineteenth century its history may be traced back to the pathologizing and medicalizing of the female orgasm. In Europe in 1653, at the time when *Sodom* was written, Pieter van Foreest notes that "in the Western medical tradition genital massage to orgasm by a physician or midwife was a standard treatment for hysteria, an ailment considered common and chronic in women." See Rachel Maines, *The Technology of Orgasm: "Hysteria," the Vibrator, and Women's Sexual Satisfaction* (Baltimore & London: The Johns Hopkins University Press, 1999), 1. At least since Greek medicine, male doctors speculated that women deprived of sexual intercourse would suffer from hysteria, a "disease" curable by massaging female genitals and inducing an orgasm. Because male penetration and orgasm were, in the androcentric view on sexuality, the only standard of sexual normality, those women who could not achieve an orgasm through "regular" intercourse were regarded as ill. The numbers were staggering. According to Thomas Syndeham, in the seventeenth century, hysteria was "the most common of all diseases except fevers" (Maines, 5). Rachel Maines insists that since doctors inherited the task of producing orgasm in more than a half of the female population and because this was a lucrative business, they were forced by necessity to combine profit with boredom and therefore invented the vibrator. What took an hour of a doctor's precious time now could be achieved in a ten minutes with a vibrator. Maines writes: "And since mechanical and electromechanical devices could produce multiple orgasms in women in a relatively short period, innovations in the instrumentation of massage permitted women a richer exploration of their physiological powers" (Maines, 11). She then concludes that "even women with very high orgasmic thresholds will usually respond eventually to vibratory massage. Those with lower thresholds can use the machine to explore their full orgasmic potential with very little fatigue" (Maines, 122). Only in the middle of the twentieth century after the medical establishment excluded hysteria from the list of illnesses and recognized women's sexuality as being independent from men's did doctors give women a device to satisfy sexual pleasures by transgressing the boundaries of "natural sexuality." Officinas's reflection that "Art should exceed what nature gave to man," has, as a corollary, the notion that mechanics expands pleasure and controls sexual identity at the same time.

    13. Steven Shapin and Simon Schaffer, *Leviathan and the Air-Pump: Hobbes, Boyle, and the Experimental Life* (Princeton: Princeton University Press, 1985), 55.

    14. Elizabeth Potter, *Gender and Boyle's Law of Gases* (Bloomington: Indiana University Press, 2001).

    15. Previously I argued that the logic of Cartesian mechanics served as a means for understanding, explaining, and inventing "sex"; here, in the case of Bacon and Boyle, I argue that the logic of "sex" served to define the scientific method of "new mechanics." Central to this transfer of "sex" into knowledge is Bacon's metaphor of scientific method as heterosexual penetration. But first, we should ask: "What is the relation between a metaphor of science and a scientific method?"

    Hellegers observes that cultural critics of science emphasize the constitutive role of representation. Scientific claims, she explains, are not independent from the general economy of representation within which analogies, metaphors, and models shape the natural world. Hellegers casts metaphors into relief by suggesting that they "do not simply describe but also prescribe and *proscribe* particular interventions in and relationships to the natural world, and to the bodies of men, women, and children." See Desiree Hellegers, *Handmaid to Divinity: Natural Philosophy, Poetry, and Gender in Seventeenth-Century England* (Norman: University of Oklahoma Press, 2000), 5. Because scientists do not have unmediated access to the natural world, scientific narrative incorporates new metaphors and may induce paradigmatic shifts, weakening the old metaphor and empowering the new. This is evidenced in the case of the "one-sex model" with its anatomical dimorphism that dominated Western thinking. The infamous seventeenth-century sexual manual *Aristotle's Master-piece,* for example, explains women as "men turn'd outside in," and men as "women with their inside out."

See Roy Porter, "'The Secrets of Generation Display'd': *Aristotle's Master-piece* in Eighteenth-Century England," in *Unauthorized Sexual Behavior During the Enlightenment,* ed. Robert Maccubbin, a special issue of *Eighteenth-Century Life: The College of William & Mary* 9, n.s. (3 May 1985): 15. Correspondingly, hermeneutic science represented through the Renaissance the relation between science and nature as the male/female unity in procreation. At the end of the Renaissance, however, a two-sex model was introduced that, according to Thomas Laqueur, took over the one-sex model in the eighteenth century. See Laqueur, *Making Sex: Body and Gender from Greeks to Freud* (Cambridge: Harvard University Press, 1990). The paradigmatic shift induced by Galilean science paralleled the shift in sexual metaphors of science. The two-sex model, with a dominating male and a subordinate female, became a constitutive model of Bacon's empirical science but also, in its hierarchical division between fallen women and redemptive men, provided the discursive conditions existed for the reconstruction of postlapsarian languages. The art of experimentation, which re-enacts the "Adamic coincidence of power and knowledge," is understood in Baconian science as a morally redeeming gesture only insofar as its metaphoric meaning invokes an explicit image of heterosexual penetration.

Here we have an epistemic and erotic meaning exchanged through the use of a metaphor of science. I am not arguing that pornography invented the sexual metaphors of seventeenth-century science; rather that the metaphor itself, to the extent that it used penetration both as pleasure and as a mechanical, rational, and punitive operation against nature and woman, was pornographic. Experimental science and pornography, in other words, stand in a constitutive and mutually supportive relationship of exclusion. Like the biblical "swine" into which demons were channeled, so too were unchaste women, at least in the case of Boyle, made ready to be demonized by sensuality and sexuality all in order to redeem men from the sins of a fallen nature. That seventeenth-century science came out of a particular and paradoxical "alignment of politics and learning" through sexuality is not peculiar only to the Italian Renaissance, testifies to the case of Bacon's experimental science. Feyerabend's lucid observation of the strange mixture of conservative and progressive elements in British empiricism seems to be revealed once again in this paradox of the simultaneous desexualization and resexualization of science.

Let us compare Aretino's erotic poetry and Baconian scientific rationality in order to demonstrate this point. One of the literary events that shook up the Italian public scene was Aretino's publication of *I modi* (sixteen positions of love-making), the crafted sonnets about each position illustrated by Gulio Romano (1499–1546), a student of Raphael, and later engraved by Marcantonio Raimondi. Aretino's sonnets, Findlen states, reconstructed the erotic logic of the erotic visual order to allow, in Aretino's own words, "the eyes to gaze at the things they must delight to see" (as quoted in Lynne Lawner, *I modi,* 9). In one of his sonnets Aretino writes:

> Open your thighs so I can look straight
> At your beautiful ass and cunt in my face,
> An ass equal to paradise in its enjoyment,
> A cunt that melts hearts through the kidneys.
> While I compare these things,

> Suddenly I long to kiss you,
> And I seem to myself more handsome than Narcissus
> In the mirror that keeps my prick erect.
> (Findlen, "Humanism, Politics and Pornography," 69)

There is a way to read Aretino not only as a precursor of a libertine literature and the Marquis de Sade, but also as a proto-experimentalist of a kind one finds in Bacon's metaphoric conception of the new empirical science. In contrast to Castiglione's purified image of the soul, Findlen claims, for Aretino "seeing was an act of (male) sexual power" (Findlen, 69), a virtue that Bacon elevates above emasculated alchemy and magic, feminized forms of knowledge. Aretino and Bacon not only shared the view of a female body holding the secrets of nature; more importantly they both believed that penetration is the investigative technique by which these secrets are revealed.

Let me press this parallel further. Marked by this metaphor Aretino and Bacon are both "expert Minister[s] of Nature" who "encounter Matter by main force . . ." (Hellegers, *Handmaid to Divinity*, 53). Because the postlapsarian language uses a metaphor of a woman to explain fallen nature and nature to explain the fall of woman, or just as "Matter" is used as a synonym for both, Bacon's and Aretino's "penetration" overlaps in meaning. Aretino's masculine analytical gaze at the woman symbolically achieves the postlapsarian recovery of the "knowledge and power in the garden of Eden" (Bono, *The Word of God and the Languages of Man*, 236) by subordinating, in a Baconian sense, the fallen feminine nature to the masculine power of "mental penetration." While the discourse of Baconian science conceals the pleasure of masculine self-eroticism, it surfaces in Aretino's lyrics, "I compare theses things . . . and I seem to myself more handsome than Narcissus." On the basis of this pleasure, the Baconian unity of "power and knowledge" over the object of domination extended the metaphor of penetration into a scientific method. The point here is not to reduce Bacon to "vulgar Freudianism" but, as Markley explains, to acknowledge a complex "overlay of the emerging discourse of gender, science, and technology" (Robert Markley, *Fallen Languages: Crises of Representation in Newtonian England, 1660–1740* [Ithaca: Cornell University Press, 1993], 226) and to put the logic of experiment under the sign of "sex." If put under this sign, Bacon's experiments or Boyle's pump work are in a sexualized space just as Aretino's imagination. Aretino too, analyzes nature by women "laid on by Manacles, that is, by extremities" (Hellegers, *Handmaid to Divinity*, 53) and brings distant descriptions of "virtual witnessing" in logical relation through the sign of sex. Male testicles, we may argue on behalf of this logic, are for Aretino "witnesses of every fucking pleasure," and are privileged by their anatomical location to witness, like Boyle's gentleman, the work of a masculine penetration of nature.

16. Max Horkheimer and Theodor W. Adorno, *Dialectics of Enlightenment*, trans. John Cumming (London: Allen Lane, 1973.) 96. See also, Immanuel Kant, *Critique of Pure Reason*, trans. (from German to English) Norman Kemp Smith (New York: St Martin's Press, 1965), A711/B739, A713/B741; Andrew Cutrofello, *Discipline and Critique: Kant, Poststructuralism, and the Problem of Resistance* (Albany: State University of New York Press, 1994), 39–41.

17. Marquis de Sade, *The Marquis de Sade. Three Complete Novels: Justine, Philosophy in the Bedroom, Eugenie de Franval and Other Writings*, compiled and trans. Richard Seaver and Austryn Wainhouse (New York: Grove Press, 1966), 68.

18. Sade, *The Marquis*, 69.

19. Maurice Blanchot, "Sade," in Marquis de Sade, *The Marquis*, 67.

20. Michel Foucault, *The History of Sexuality Volume 1: An Introduction*, translated from French by Robert Hurly (New York: Vintage Books, 1990), p. 71.

21. Horkheimer and Adorno bring to our attention Sade's subversive civilizational project, which represents the new economy of pleasure in the Age of Reason. They point out that Sade uses the character Juliette and the "intellectual pleasure" operating in his sexual laboratories as *"amor intellectualis diaboli"* to embody "the pleasure of attacking civilization with its own weapons" (Horkheimer and Adorno, *Dialectics of Enlightenment*, 94).

22. Jacques Lacan, "Kant with Sade," trans. James B. Swenson, Jr., October 51 (Winter 1989): 55.

23. See Michel Foucault's introduction to Georges Canguilhem, *The Normal and the Pathological*, trans. Carolyn R. Fawcett in collaboration with Robert S. Cohen (New York: Zone Books, 1991), 14; Thomas Kuhn, "Alexandre Koyré and the History of Science. On an Intellectual Revolution," *Encounter* 34, no. 1 (January 1970): 67–69.

24. Salviati talking to Sagredo from Galileo Galilei, *Dialogues Concerning the Two Great Systems of the World*, trans. Thomas Salusbury (London: 1661), 301, #355, as quoted by Edwin Arthur Burtt, *The Metaphysical Foundations of Modern Physical Science: A Historical and Critical Essay* (London: Routledge and Kegan Paul, 1967), 69. On this point also, see Susan R. Bordo, *The Flight to Objectivity: Essays on Cartesianism and Culture* (Albany: State University of New York Press, 1987), 34; see also Alexander Koyré, *From the Closed World to the Infinite Universe* (Baltimore: John Hopkins Press, 1957), 90.

25. This politics of shame in writing history has been disguised in William Lecky's history of European morals. At the point at which he had to introduce chastity and its role in this history he felt the need to apologize, "I am sorry to bring such subject before the reader." See Peter Brown, *The Body and Society: Men, Women, and Sexuality Renunciation in Early Christianity* (New York: Columbia University Press, 1988), xvi.

26. Alexandre Koyré, *Galileo's Studies* (New Jersey: Humanities Press, 1978), 37.

27. Alexandre Koyré, *Metaphysics and Measurement: Essays in Scientific Revolution* (Cambridge: Harvard University Press, 1968), 13–14.

28. Koyré stresses that the power of pure mathematical rationality over the bodily experiences leads Galileo to his most important discoveries. For instance, to claim that a heavier body will fall faster than a lighter body is to be consistent with our observation but not with mathematical imagination according to which all falling bodies have equal acceleration. It is counter-experiential to claim, as Galileo did in contrast to Aristotle, that only time, rather than weight or height, has a relationship to free fall. Koyré claims that the latter understanding in particular depended on a geometrical imagination. If one imagines speed as a continuous passage, as did Galileo, then motion can be represented by an isosceles triangle: one side represents speed while the second represents infinite time intervals; the third side of the triangle connects every point of speed with any point in time (Koyré, *Galileo's*, 105). The triangle, rather than Galileo's senses, he continues, allowed Galileo to show that degrees of speed corre-

spond to an infinite number of time units. By tying speed in his imagination to an abstract and mathematical concept of time, and by abstaining from his sense experience in favor of a triangle, Galileo realized that speed, like time, can be represented by an infinite number series. Furthermore, Koyré insists, Galileo realized that the order of nature reveals itself through discourse rather than the senses. If time intervals can be endlessly divided and subdivided and if they always correspond to certain degrees of speed, it follows that speed is infinite and thus permanent. Speed, then, is not uniform—as Aristotelians held—but the acceleration of falling objects is. Just as uniform motion can be described as equal distances traversed in equal times, so can uniform acceleration be explained as equal increases of speed in equal times. Thus, Koyré concludes, Galileo formulates correctly on the grounds of a mathematical construction of time that in equal intervals of time—counting from the time of rest— speed will double in the second equal time interval, triple in the third, quadruple in the fourth, and so on. Since time was not an empirical, experiential, intuitive category, but rather, a mathematically constructed one, time became a measurable and therefore *objective* variable.

29. Koyré, *Metaphysics*, 80.

30. Plato, *Symposium* (New York: Liberal Arts Press, 1956), 211e; also, see Evelyn Fox-Keller, *Reflections on Gender and Science* (New Haven: Yale University Press, 1985), 21–32; Luce Irigaray, *An Ethics of Sexual Difference,* trans. Carolyn Burke and Gillian C. Gill (Ithaca: Cornell University Press), 20–33.

31. John M. Dillon, "Rejecting the Body, Refining the Body: Some Remarks on the Development of Platonist Asceticism," *Asceticism,* ed. Wimbush and Valantasis (Oxford: Oxford University Press, 1995), 86f.

32. 1.1.1214a 16ff.

33. 10.9.1179b20ff.

34. Foucault, *The Use of Pleasure: The History of Sexuality*, vol. 2, trans. Robert Hurley (New York: Vintage Books, 1990), 72.

35. Ibid.

36. Ibid., 64 and 78.

37. Lawrence D. Kritzman, "Foucault and the Ethics of Sexuality," *L'Espirit createur* 25, no. 2 (Summer 1985): 86–96; Ladelle McWhorter, "Asceticism/*Askēsis:* Foucault's Thinking Historical Subjectivity," in *Ethics and Danger,* ed. Arleen B. Dalley (Albany: State University of New York Press, 1992), 243–54; Geoffrey Galt Harpham, "Foucault and the 'Ethics' of Power," in *Ethics/Aesthetics: Post-Modern Positions,* ed. Robert Merrill (Washington, D.C.: Maisonneuve Press, 1988), 71–81; John Rajchman, "Ethics after Foucault," *Social Tex* 13, no. 14. (Winter/Spring 1986): 165–83; David M. Halperin, *Saint Foucault.*

38. Foucault, *Language, Counter-Memory, Practice* (Ithaca, N.Y.: Cornell University Press, 1977), 202.

39. Ibid.

40. Dillon, "Rejecting," 80–87; Kallistos Ware, "The Way of the Ascetics: Negative or Affirmative?," in *Asceticism,* ed. Wimbush and Valantasis (Oxford: Oxford University Press, 1995), 3–15; Edith Wyschogrod, "The Howl of Oedipus, The Cry of Heloise: From Asceticism to Postmodern Ethics," in *Asceticism,* ed. Wimbush and Valantasis (Oxford: Oxford University Press, 1995), 16–30; Bruce J. Malina, "Paine,

Power, and Personhood: Ascetic Behavior in the Ancient Mediterranean," in *Asceticism,* ed. Wimbush and Valantasis (Oxford: Oxford University Press, 1995), 162–77; J. Gilles Milhaven, "Asceticism and the Moral Good: A Tale of Two Pleasures," in *Asceticism,* ed. Wimbush and Valantasis (Oxford: Oxford University Press, 1995), 375–94; Verna E. F. Harrison, "The Allegorization of Gender: Plato and Philo on Spiritual Childbearing," in *Asceticism,* ed. Wimbush and Valantasis (Oxford: Oxford University Press, 1995), 520–43; Geoffrey Galt Harpham, *The Ascetic Imperative in Culture and Criticism* (Chicago: The University of Chicago Press, 1993); Brown, *The Body and Society;* Bruce S. Thornton, *Eros: The Myth of Ancient Greek Sexuality* (Boulder, Colo.: Westview Press, 1997); Pierre Hadot, *Philosophy as a Way of Life: Spiritual Exercises from Socrates to Foucault,* trans. Michael Chase (Oxford: Blackwell, 1995).

41. Quoted in Fox-Keller, *Reflections,* 52.

42. Herbert Marcuse, *Eros and Civilization: A Philosophical Inquiry into Freud* (Boston: Beacon Press, 1966), 171.

43. Giovanni Dall'Orto, "'Socratic Love' as a Disguise from Same-Sex Love in the Italian Renaissance," *Journal of Homosexuality* 16 (1989): 33–65.

44. Ibid., 39.

45. Ibid.

46. Ibid., 40.

47. Ibid., 88.

48. Findlen, "Humanism, Politics and Pornography," 87.

49. Lawren, "Introduction," 45.

50. Findlen, "Humanism, Politics and Pornography," 93

51. Ibid., 92.

52. Ibid., 89

53. Barkan, *Transuming Passion,* 72.

54. In Findlen, "Humanism, Politics and Pornography," 94.

55. Ibid.

56. Findlen, "Humanism, Politics and Pornography," 59, 63.

57. Dominique Dubarle, "Galileo's Methodology of Natural Sciences," in *Galileo Man of Science,* ed. Ernan McMullin (Princeton, N.J.: The Scholar's Bookshelf, 1988) 306.

58. Stillman Drake, *Galileo at Work* (Chicago: University of Chicago Press, 1981) 73; Galileo Galilei, *Dialogues Concerning the Two Chief World Systems—Plolomeic and Copernicon,* trans. Stillman Drake (Berkeley: University of California Press, 1967), 227; *Opere* 8, 277.

59. Dubarle, "Galileo's Methodology," 306.

60. Steven Shapin and Simon Schaffer, *Leviathan and the Air-Pump: Hobbes, Boyle, and the Experimental Life* (Princeton: Princeton University Press, 1985), 55–60.

61. Markley, *Fallen Languages,* 221.

62. Lawner, Introduction to *I modi,* 44.

63. Guido Ruggiero, *The Boundaries of Eros: Sex Crime and Sexuality in Renaissance Venice* (Oxford: Oxford University Press, 1989), 146–47.

64. Theodore Porter, *The Trust in Numbers: The Pursuit of Objectivity in Science and Public Life* (Princeton: Princeton University Press, 1995), 3.

65. Mario Biagioli, "Knowledge, Freedom, and Brotherly Love: Homosociality and the Accademia dei Lincei," *Configurations,* 1995, 2, p. 141.

66. David Noble, *The World Without Women, The Christian Clerical Culture of Western Science* (Oxford: Oxford University Press), 216–17.

67. From the standpoint of Foucault's history of sexuality Neoplatonism carried on the sexual politics of heterosexual principles of pleasure initiated strongly by the Church fathers around the fifth and the sixth centuries. The case of the conflict between English hermetic philosophy and "new mechanics" is a prime example of the radicalization of masculine sexuality through science. By contrast to "new mechanics" which, following Neoplatonic dualism, aspired to sequester mind from matter, for hermetic philosophy the spirit dwells in matter, allowing sensual experience to be touched by truth. This union between mind and matter hermetic philosophers represented with a metaphor of sexual union between male and female, mind and nature. Thomas Vaughan, proponent of hermeneutic philosophy, articulates science as conjugal relation as follows:

> As . . . the conjunction of male and female tends towards a fruit and propagation becoming the nature of each, so in man himself that interior and secret association of male and female, to wit the copulation of male and female soul, is appointed for the production of fitting fruit of Divine Life. . . . Marriage is a comment on life, a mere hieroglyph or outward representation of our inward vital composition. For life is nothing else but a union of male and female principles, and he that perfectly knows this secret knows the mysteries of marriage—both spiritual and natural. . . . Marriage is no ordinary trivial business, but in a moderate sense sacramental. It is a visible sign of our invisible union with Christ. (Quoted in Fox-Keller, *Reflections,* 50)

This metaphor of science as sexual union between male and female to a large extent determines the "secret" character of scientific practices and findings. Just as the pleasure of marital sex remains intimate and private in order to protectively surround the sacred process of impregnation and procreation to occur, so the alchemist intimates male and female union in his laboratory. Only through this intimate participation between male and female does nature reveal itself in the alchemist laboratory. The metaphor of sexual intimacy also presents knowledge as an intimate and tacit component of this union. Perhaps this explains why hermetic knowledge was not made explicit and thus could not satisfy the new criteria of objective verification based on the chaste sexual metaphor of science.

Including, at least allegorically, the feminine principle in science did not make alchemists precursors of feminism, but formulated the underlying economy of pleasure as procreative sexuality. If in this economy of pleasure hermetic philosophers found their knowledge, "new mechanics" found it in the Neoplatonic model of chastity. Whereas the alchemist desire for procreation included positive pleasures of the body, mechanical philosophers insisted on asceticism, or negating bodily pleasures.

Although in conflict, hermetic and new mechanical philosophy still shared in common the same principles of sexual divisions, exclusions, and subordination of women. Philosophers such as Francis Bacon cherished this principle of exclusion along gender lines as a necessary reform of science. In his early work, *The Masculine*

Birth of Time, he explains: "Older science," he claims, "represented only a female offspring, passive, weak, expectant." *The Masculine Birth of Time,* in Fox-Keller, *Reflections,* 38. Because a woman subordinates sober inquiries to her seductive sentiments, which by its nature is "ungodly, even Satanic" (Ibid., 59). Bacon thus prescribes a "chaste and lawful marriage between Mind and Nature" that will "bind [Nature] to [man's] service and make her [his] slave" (Ibid., 48). He also recommends chastity as a method of cleansing philosophers' minds: "Men are to be entreated again and again. . . . That they should humbly and with a certain reverence draw near to the book of Creation. . . . That on it they should mediate, and that then washed and clean they should in chastity and integrity turn them from opinion" (Ibid., 38n). Bacon clearly calls for the exclusion of the feminine and the sexual from new science in order to make it a male and chaste endeavor.

Mechanics, in the view of Bacon, More, and Glanvill, became a domain of exercising moral purity in opposition to the alchemical fusion of sexuality and knowledge which, in their views, encroached upon science as a fire from hell. Baconian new science, the formation of the Royal Society, and political conservatism consolidated the new economy of homosocial pleasures and power/knowledge around male chastity. As Evelyn Fox-Keller claims, the new science also became a new model of universal domination and control through the chaste rationality. The epistemology of the chaste body only metaphorically remains detached from the world it explains; on a much deeper level, precisely because of its chastity, this epistemology never stops sexualizing the world in order to maintain its own purity.

68. Alexandre Koyré, *From the Closed World to the Infinite Universe* (Baltimore: Johns Hopkins University Press, 1975), 90; Andrew Weeks, *Boehme: An Intellectual Biography of the Seventeenth-Century Philosopher and Mystic* (Albany: State University of New York Press, 1991), 148.

69. Foucault, Introduction to Georges Canguilhem, *The Normal and the Pathological,* trans. Carolyn R. Fawcett in collaboration with Robert S. Cohen (New York: Zone Books, 1991), 13.

70. Foucault, "Introduction," 14.

71. Ernst Mach, *The Science of Mechanics: A Critical and Historical Account of Its Development,* trans. Thomas J. McCormack (La Salle, Ill.: The Open Court Publishing Co., 1942), 168.

72. Sade, *Juliette,* trans. Austyn Wainehouse (New York: Grove Press, 1968), 584.

73. Carolyn Merchant, *The Death of Nature: Women, Ecology, and the Scientific Revolution* (San Francisco: Harper & Row, 1983), 172.

74. Leo Bersani, *The Freudian Body: Psychoanalysis and Art* (New York: Columbia University Press, 1986), 54.

75. Marquis de Sade, *120 Days of Sodomy and Other Writings,* compiled and trans. Austryn Wainhouse and Richard Seaver (New York: Grove Press, 1966), 392–93 and 602.

76. Marquis de Sade, *Philosophy in the Bedroom,* 271 and 365.

77. Sade, *The 120 Days of Sodom,* 670–72.

78. Marcel Henaff, "Sade, the Mechanization of the Libertine Body, and the Crisis of Reason," in *Technology and the Politics of Knowledge,* ed. A. Feenberg, and A. Hannay (Bloomington: Indiana University Press, 1995), 219–20.

79. About the voyeurism of the scientific gaze see an illuminating article by Barbara Maria Stafford, "Voyeur or Observer? Enlightenment Thoughts on the Dilemmas of Display," *Configurations* 1, no. 1 (Winter 1993): 95–128.

80. Foucault, *The History,* 21.

81. Henaff, "Sade," 213–35.

82. Ibid.

83. Ibid., 229.

84. But is rational discourse eroticized only because it represents objectively sexual acts, or is there an intrinsic pleasure in the rational analysis itself? Roland Barthes comes in on the side of the latter interpretation. Sade, he claims, invented a new form of discursive pleasure, a superior pleasure, with a mathematical idea. Barthes argues: "Sadian debauchery, usually referred to only as a function of the philosophical system of which it is no more than the abstract cipher, in fact participates in an *art de vivre:* in it is inscribed the concomitance of the pleasures." Roland Barthes, *Sade, Fourier, Loyola,* trans. Richard Miller (New York: Hill and Wang, 1976), 142.

85. Jean Baudrillard, *Forget Foucault & Forget Baudrillard* (New York: Semiotext(e) Foreign Agents Series, 1987), 33.

86. Ibid., 29.

87. Ibid., 31

88. Bersani, *The Freudian Body,* 52.

89. Gilles Deleuze describes this aspect of Sade's narrative:

The descriptions, the attitudes of the bodies, are merely living diagrams illustrating the abominable descriptions; similarly the imperatives uttered by the libertines are like the statements of problems referring back to the more fundamental chain of sadistic theorems: "I have demonstrated it theoretically," says Noirceuil, "let us now put it to the test of practice."

We have therefore to distinguish two factors constituting a dual language. The first, the imperative and descriptive factor, represents the *personal* element; it directs and describes the personal violence of the sadist as well as his individual tastes; the second and higher factor represents the *impersonal* element in sadism and identifies the impersonal violence with an Idea of pure reason, with a terrifying demonstration capable of subordinating the first element. In Sade we discover a surprising affinity with Spinoza—a naturalistic and mechanistic approach imbued with mathematical spirit. This accounts for the endless repetitions, the reiterated quantitative process of multiplying illustrations and adding victim upon victim, again and again retracing the thousand circles of an irreducibly solitary argument. (Gilles Deleuze, *Masochism: An Interpretation of Coldness and Cruelty,* trans. Jean McNeil and Aude Willm (New York: Zone Books, 1991), 19–20)

90. Ibid.

91. Deleuze writes:

It is remarkable that the process of desexualization is even more pronounced than in neurosis and sublimation; it operates with extraordinary coldness; but it is accompanied by a resexualization which does not in any way cancel out the desexualization, since it operates in a new dimension which is equally remote from functional disturbances and from sublimations; it is as if the desexualized element were resexualized but nevertheless retained, in a different form; the desexualized has become in itself the object of sexualization. This explains why coldness is the essential feature of the structure of perversion; it is present both in the apathy of the sadist, where it figures as theory, and in the ideal of the masochist, where it figures as fantasy. The deeper the coldness of the desexualization, the more powerful and extensive the process of perverse resexualization; hence we cannot define perversion in terms of a mere failure of integration. Sade tried to demonstrate that no passion, whether it be political ambition, avariciousness, etc., is free from "lust"—not that lust is their mainspring but rather that it arises at their culmination, when it becomes the agent of their instantaneous resexualization. (*Masochism,* 117–18)

92. Ibid.
93. Ibid.
94. Henaff, *Sade,* 41.
95. Jonathan Dollimor, *Sexual Dissidence: Augustine to Wilde, Freud to Foucault* (Oxford: Claredon Press, 1991), 324.
96. Michel Foucault, *Language, Counter-Memory, Practice,* 151.
97. Friedrich Nietzsche, *Basic Writings of Nietzsche,* trans. and ed. Walter Kaufman (New York: The Modern Library, 1968), 589.

CHAPTER 3. THE JESUITS' HOMOSOCIAL TIES AND THE EXPERIMENTS WITH GALILEO'S PENDULUM

1. William J. Summers, "The Jesuits in Manila, 1581–1621: The Role of Music in Rite, Ritual, and Spectacle," *The Jesuits: Cultures, Sciences, and The Arts, 1540–1773,* ed. John W. O'Malley (Toronto: University of Toronto Press, 1999), 661.
2. Peter Dear, *Discipline and Experience: The Mathematical Way in the Scientific Revolution* (Chicago: University of Chicago Press, 1995), 9.
3. Rivka Feldhay, *Galileo and the Church: Political Inquisition or Critical Dialogue?* (Cambridge: Cambridge University Press, 1995), 125.
4. Feldhay, *Galileo and the Church,* 110–98, and 73–92; see also Jerome J. Langford, *Galileo, Science and the Church* (Ann Arbor: University of Michigan Press, 1992), 11; Ann W. Ramsey, "Flagellation and the French Counter-Reformation: Asceticism, Social Discipline, and the Evolution of a Penitential Culture," in *Asceticism,* ed. Wimbush and Valantasis (Oxford: Oxford University Press, 1995), 576.
5. But this ascetic discipline raised some questions about the appropriateness of male teachers and students living together. M. Laurentano, head of the German College, explains this as the contradiction between secular education and the requirement of the monastic ideal of "otherworldliness." He wrote: "That religious people like us should become educators of secular youth seems inappropriate. It requires us to leave our residence, to eat and live with them, which is not without a certain danger, or at least a suspicion of the people of ['this'] world" (in Feldhay, *Galileo and the Church,* 120).

6. Ibid., 198.
7. Ibid., 124–25.
8. Although the Jesuits did not create a scientific revolution single-handedly, they unquestionably created a scientific *pedagogy*. Here is a brief outline of the history of their education. The following should provide ample evidence to support my claim that the Jesuits, though in the vanguard of mathematical physics, had still carefully related the body to a scientific rationality. The Jesuit capacity for teaching science surpassed any other group at that time. The Academy in Lincei, as well as Leibniz's plan for a German academy of sciences, was modeled on the Jesuits's "quasi-monastic" orientation. See J. L. Heilbron, *Electricity in the 17th and 18th Centuries: A Study of Early Modern Physics* (Berkeley: University of California Press, 1979), 107; also, David Noble, *The World Without Women, The Christian Clerical Culture of Western Science* (Oxford: Oxford University Press), 214. According to Langford, the Protestant revolt had a harmful effect on education, rapidly decreasing the numbers of students. Funds set aside by the Catholic Church for schools were confiscated by Protestant authorities, and many secondary schools run by monasteries were closed. In England, Langford claims, the numbers of students decreased by half at the end of the sixteenth century (*Galileo, Science and the Church,* 3–4). Slashes in funding in Protestant countries hindered scientific education based on demonstrational and instrumental pedagogy. In addition to having more funds, Jesuit schools were well-equipped with scientific instruments because they preferred hands-on knowledge, observations, and exercises as a part of learning. Perhaps this may explain the envy an Englishman expressed to his friend after his journey abroad: "This I must always affirm for the honor of my mother the University of Oxford, that if her children had the good utensils, which adorn the colleges of the Jesuits abroad, the world would not long want good proof of their ingenuity" (in Heilbron, *Electricity,* 103). Noah Porter observes, "The Jesuit institutions are not limited in the material of instruction. Money, buildings, apparatus, and libraries are supplied in sufficient abundance" (*The Educational Systems of the Puritans and Jesuits Compared* [New York: M. W. Dodd, 1851], 56). The revised *Ration* of 1832 sanctions into a rule an educational abundance of scientific equipment that was already the established tradition in the Jesuit schools: "The same is to be said of literary periodicals for the use of the Professors; of museums, physical apparatuses, and other equipments, which are needed by a college according to its degree" (Thomas Hughes, *Loyola and the Educational System of the Jesuits* [New York: Charles Scribner's Sons, 1892], 187). This commitment to scientific instrumentation is strictly linked to the Jesuit commitment to observation. For example, the Order built numerous observatories specifically for pedagogical purposes. Again, the Reverend Hughes testifies:

> Father Hubert superintended the building of an observatory at Wurzburg; Father Maximilian Hell, the court astronomer, built one at Vienna. At Mannheim, a third was erected by Mayer and Metzger; at Tyrnau, one by Keri; at Prague, another by Stepling; one at the Jesuit College of Graz; similarly at Wilna, Milan, Florence, Parma, Venice, Brescia, Rome, Lisbon, Marseilles, Bonfa. In short, Montcula remarks: "In Germany and the neighboring countries, there were few Jesuit colleges without an observatory." (Hughes, *Loyola,* 172)

Vignali, Antonio: author of *La Cazzaria*, 44–45
Virilio, Paul, 31
virtual witnessing, 11; and sex, 46–48, 174n. *See also* Boyle, Robert

Viviani, Vincenzo, 196n

Wittgenstein, Ludwig, 85

Xenophon, 22

renaissance: and debate on love, 43; Italian, erotic literature of, 35, 44–45; and relation between science and nature, 173n; and sexuality, 47–48
res cogitans, 54, 170n
res extensa, 170n
resexualization, 2, 19, 21, 31, 55, 155, 173n, 180n–181n
Riccioli, Giovanni Battista: experiments with pendulum, 59, 65, 70–73, 75, 76, 149, 154–55. *See also* Galilei, Galileo; Jesuits; pendulum
Rocco, Antonio, 44, 45, 155
Rochester, John Wilmot, 2nd Earl of, 171n
Romano, Giulio, 173n. *See also I Modi*
Roothaan, Jan Philip: General of Society of Jesus, 184n, 189n. *See also* Jesuits
Royal Society of London, 64, 179n
Rubin, Gayle: and fisting, 171n

Sade, Marquis de, 14, 175n, 180n–181n; and apathy, 37, 55; and Galilization of sexuality, 53–54; and *120 Days of Sodom,* 52; and *Philosophy in the Bedroom,* 52; and "objectivity" as pleasure, 51–57
sadism, 180n–181n. *See also* Sade, Marquis de
Schaffer, Simon, 36, 46, 196n
Schott, Robin May: and *Cognition and Eros,* 33
Sedgwick, Eve Kosovsky, 66. *See also* homosociality
Serres, Michel: "Abriosia and Gold," 158
Settle, Thomas, 130
seventeenth-century science: 46, 153, 156, 186n; concepts of objectivity in, 33–38; sexual metaphors of, 173n, 178n–179n. *See also* Boyle, Robert; Jesuits; Koyre, Alexandre; neoplatonism; pendulum; Sade, Marquis de
sex: as "coded operation", 30; instrumentation of, 50, 54. 171n; and mathematics, 51–55; and mechanics, 30, 31, 50, 51, 170–171n; medicalization of, 171–172n; as representational sign of body-instrument link, 76. *See also* pornography; Rochester, Earl of; Sade, Marquis de; "virtual witnessing"
Shapin, Steven, 33–34, 36, 46. *See also* Boyle, Robert
Sheele, Sister M. Augustine, 183n
Sherman, Stuart, 20, 27
Singer, Peter, 2
Socrates, 22. 42
Sodom: a Play, 171n. *See also* Rochester, Earl of
sodomy, 43, 47–48. See also *La Cazzaria, I Modi, Sonnetti lussuriosi*
Solomon, Julie Robin, 169n
Sonnetti lussuriosi, 44. *See also* Aretino, Pietro
speed: and pleasure, 31; and mathematics, 176n
standpoint: in Galilean physics, 142–43
Stone, Rosanne Allucquere: and *The War of Desire and Technology at the Close of the Mechanical Age,* 30
synchrony. *See* pendulum
Syndeham, Thomas, 172n

technologies of self, 159, 187n
technophilia, 30
telesexuality, 31
Thérèse Philosophe, 170n–171n
time: ancient Greek concept of, 21–22; and bodily discipline, 23–24; in Christian eschatology, 22–23; eroticization of, 21. *See also* mechanical time
Trebizond, George of, 43. *See also* Ficinio, Marsilio; Plato
two-sex model, 34, 173n, 178n

Vaughan, Thomas, 178n
Vermeer, Jan: *Woman Holding a Balance,* 158
vibrator; history of, 171n–72n. *See also* orgasm; sex, medicalization of

Index

neoplatonism: and discourse on sexuality, 7, 14, 38; and exclusion of the feminine, 178n–179n; and influence on Galileo, 38; and science, 42–43, 46–47; and voyeuristic gaze, 35. *See also Accademia dei Lincei*; Jesuits; Koyre, Alexandre; pornography; Renaissance; two-sex model

Newton, Isaac, 8, 68

Nietzsche, Friedrich, 2, 9, 56, 81; and notion of creative value, 162n

Noble, David, 48

nomos, 22. *See also* time, ancient Greek concept of

North, J.D., 25

"object" explicator, 149

one-sex model, 173n; as metaphor of science in alchemy, 178n

orgasm, 31, 171n–172n

pantograph: invented by Jesuits, 183n

pedagogy, 77; and "Agnes", 82; defined as "knowing with an eroticized body," 81; of practices, 13–14, 86. *See also* Jesuits, secular education (pedagogy)

pendulum: as discourse on sexuality, 5, 8; Galileo's specifications for construction, 135–36; geometric theorems associated with, 122–28, 194n; as instrument of time measurement, 14, 19, 21, 40, 70, 148; *Lebenswelt* structure of approach to; and "Livingston's pair;" 131, 148, 149; mechanics of, 19, 40, 41, 117–20 (and diagrams); operationalization of, 137–42, 157, 159; performative geometry of, 144; "received view," of, 195n–196n; types of (and diagrams), 121. *See also* clock, pendular; Huygens, Christiaan; Jesuits; Livingston, Eric; Riccioli, Giovanni Battista

personal watch: 28–29, 31; as erotica (with illustration) 28–29. *See also* clock, pendular; mechanical time

phusis, 22, 42. *See also* time, ancient Greek concept of

physics: Aristotelian, 22, 41, 120, 134; experimental, 2, 3, 63; mathematical, 5, 11, 123, 154, 166n, 182n, 185n

Plato, 2, 22; and sensual/conceptual relation, 41–43. *See also* Ficinio, Marsilio

pleasure of analysis, 2, 6, 9, 14, 40, 53, 153, 154, 162n, 180n. *See also* Foucault, Michel

pornography, 26, 155; and experimental science, 173n; and mechanics, 35–36, 50, 170n–172n; and Renaissance, 35. *See also* mathematics; mechanical time; Renaissance; Rochester, Earl of; Sade, Marquis de; sex

Porter, Noah, 67, 182n

Porter, Theodore, 33, 188n

Potter, Elizabeth, 36. *See also* Boyle, Robert

praxiom: defined, 88; and pendulum, 139, 149; and prism, 99, 101, 112

prism, 15; establishing "body-instrument link" with (instructions and diagrams), 87–88; discovering mathematical objectivity of symmetry and pattern in (instructions and diagrams), 98–101; finding reflective patterns in (instructions and diagrams) 88–92, 94–97; finding rule of reflection in (instructions and diagrams) 102–11; obtaining correct type of, 192n (with diagram, 193n), triangle as theorem-organizing object in, 107

"queer" practices, 83, 85. *See also* Foucault, Michel

Rabinow, Paul, 192n

Raimondi, Marcantonio, 173n. *See also I Modi*

reconstructivists, 10, 164n. *See also* Drake, Stillman

Jesuits (Society of Jesus), 15; asceticism of, 61, 62, 65, 68; and astronomy, 182n, 186n; and Counter-Reformation, 60–61; and Galileo, 59, 62, 65, 184n–185n; and homosocial bonding, 59–60, 66–67, 68–70, 188n; as missionary order, 186n, 190n; scientific rationality of; 59–60, 62, 63, 64–65, 67–68, 181n, 184n; and secular education (pedagogy), 61–62, 182n–184n, 188n–189n; Thomism of, 61–62. *See also* mathematics; Riccioli, Giovanni Battista

kairos, 21. *See also* time, ancient Greek concept of
Kant, Immanuel, 33; and apathy, 37; and "moral geometry," 14, 37–38
Keller, Evelyn Fox, 51, 179n
knowing-body: and pendulum, 12–13, 115, 151, 156–57; and prism, 86–87. *See also* body-instrument link
Koyre, Alexandre: historian of Galileo's Science, 10–11, 14, 49–50, 53, 63, 70, 71–73, 115, 123, 149, 175n–176n, 190n; and theory of objectivity, 38–41. *See also* reconstructivists
Kristeva, Julia: and *Desire in Language*, 74, 191n
Kuhn, Thomas, 6, 38, 115

Lacan, Jacques, 6, 38, 74
Landes, David S., 23–24, 25, 28–29
Langford, Jerome J., 182n, 184n–185n
Laqueur, Thomas, 173n
Latour, Bruno, 12
law of free fall, 115–17, 175n–176n, 197n. *See also* pendulum
law of lengths, 123–24. *See also* pendulum
Lawner, Lynne: and Introduction to *I Modi*, 35
Lecky, William, 175n
Lectio divina, 73. *See also* Loyola
Livingston, Eric, 13, 130–131, 133, 195n

Loyola, Ignatius, Saint: founder of Jesuit order (Society of Jesus), 65, 67–68; *Spiritual Exercises* of, 68, 73
Lynch, Michael, 13

Mach, Ernst, 14, 49
MacIntyre, Alasdair: and *After Virtue: A Study in Moral Theory*, 169n
MacLachlan, John, 121, 130
Maines, Rachel, 172n
Marcuse, Herbert, 10, 43
Markley, Robert, 11, 46, 174n
Marx, Karl: and "economy of time," 165n; and theory of general equivalence, 166n
Mathematics: as abstract unifier, 153–54; and Jesuits, 59, 60; 63–64, 65, 184n; and pleasure, 51–54; and prisms, 101–2. *See also* pendulum; pornography; Sade, Marquis de
mathesis universalis, 50, 54, 153
McNamara, JoAnn, 186n
McNeill, William H., 23
mechanical time, 14; and collective sexual patterns, 27–28; and diurnal prose form, 27; in industrial society, 19, 24; 26–27; and modern subjectivity, 20; in monasteries, 21, 23–26, 168n. *See also* clock, pendular; Marx, Karl; personal watch; Riccioli, Giovanni Battista; speed
Medici, 62
Merchant, Carolyn, 51
Mersenne, Marin, 34, 72, 190n
I modi, 35, 173n–174n. *See also* Aretino, Pietro; Lawner, Lynne
moral geometry. *See* Kant, Immanuel
More, Henry, 42
Mumford, Lewis, 24, 26–27. *See also* mechanical time

Nadal, H: and "Rules for Scholars of the Society," 60. *See also* Jesuits
natural philosophy, 3, 8, 34, 59, 170n, 185n, 189n
Naylor, Ronald, 190n

187n; on creative force, 2–3; on "desire," 8–9; on discipline, 191n; on discursive formations, 192n; and fisting, 171n (*See* Rubin, Gayle); on "pedagogy of practices," 86; on rules and perversion, 56–57; on Sade, 52; on sex and Christianity, 64
Fox, David, 170n
Frazee, Charles A.: historian of Roman Catholic Church, 66
Freud, Sigmund, 8, 12, 54, 153; and *Civilization and its Discontents*, 2, 8, 153

Galdikas, Birute, 2
Galilei, Galileo, 5, 6, 8, 11, 13, 14, 38, 41, 42; 48, 52; and *Accademia dei Lincei*, 49; and astronomy, 123; and *Il Saggiatore*, 184n; and inclined plane experiment, 197n; and letter to Guidobaldi, 126; and self-fashioning, 4; trial of (for Copernican views), 62, 65; and *Two World Sciences*, 15. *See also* Biagioli, Mario; Jesuits; Koyre, Alexandre; pendulum
"Galilean physics," 99, 140, 149: defined, 142; instrument use in, 141, 148–49; intersubjectivity of, 142; as pairing object, 154; reproducibility of, 143
"Galilean science," 4, 11, 133; differentiated from "Galileo's science," 163n. *See also* "handmaid;" Jesuits; seventeenth-century science
Galileo's science: and *l'absole indetermine*, 49–50; and body-related practices, 11; as erotic improvisation, 81; pedagogical structure of, 134–35; in textbooks, 135. *See also* Koyre, Alexandre; pendulum; Sade, Marquis de
Garfinkel, Harold, 12, 13, 84; and "Agnes," 81–82, 157
Geographia Reformata, 186n
geometry, 35, 40, 41, 144, 157. *See also* pendulum; prism
Glaber, Raoul, 168n

Le Goff, Jacques, 25
Gooding, David, 196n
Goux, Jean-Joseph, 20, 166n. *See also* Marx, Karl
Grienberger, Christoph: Jesuit mathematician, 185n
Grimaldi, Francesco M.: Jesuit scientist, 73, 186n

Hall, Edward T., 20
"handmaid," 8, 11, 74, 129, 156. *See also* Boyle, Robert; Hellegers, Desiree
Haraway, Donna, 2
Heilbron, J.L., 182n, 183n
Hellegers, Desiree, 11; and *Handmaid to Divinity*, 172n, 174n
Henaff, Marcel, 51, 52, 55
hermetic science, 173n. *See also* alchemy; one-sex model; two-sex model
homoeroticism, 4, 12, 43, 44–45; and Galilean scientific revolution, 49. See also *Accademia dei Lincei*; homosociality; Jesuits; seventeenth-century science
homosociality: 3, 154; and early Christian militaristic ethos, 66; and women, 69–70. *See also* Jesuits
"Hook," 11. *See also* "handmaid;" Boyle, Robert
Horkheimer, Max, 175n
Hughes, Reverend Thomas, 182n, 186n
"human clock," 75. *See also* Riccioli, Giovanni Battista
Hunter, Ian, 4
Husserl, Edmund, 4, 163n, 192n, 197n
Huygens, Christiaan, 19, 70, 194n; letter to Descartes, 183n. *See also* clock, pendular

"instructed action": and prism, 87
Irigaray, Lucy, 69, 155
isochrony. *See* pendulum

Jacob, Margaret C., 35–36, 170n. *See also* pornography

200 INDEX

Brown, Peter, 22
Butler, Judith, 3, 156, 162n. *See also* "erotic improvisation"

Calvinism, 28, 62
canonical hours, 25, 167n. *See also* mechanical time
Cassian: discussion of chastity, 187n–188n
Castiglione, Baldassare, 46, 174n
catalogue, *Nasco Science '96*: ordering pendulum components from, 129. *See also* pendulum
Catholic Church, 30, 45, 60–61, 66, 182n
La Cazzaria, 44–45. *See also* Vignali, Antonio
Cesi, Federico, 48, 49. *See also* Accademia dei Lincei
Cistercians: and bodily discipline, 25
Clavius, Christopher: Jesuit astronomer, 184n
clock, pendular, 14; as instrument of self-regulation, 19–20; invention of, 19, 70; public reverence for, 26. *See also* Huygens, Christiaan; mechanical time; pendulum; personal watch; pornography
clock, Roman water, 164n
Cluniac order: and bodily discipline, 25
Collegio Romano, 184n, 185n
contingency. *See* pendulum, operationalization of
Copernicanism, 38, 60, 61, 62, 186n
Council of Trent, 60
Counter-Reformation, 60–61, 66
Crash: novel by J.G. Ballard, 31
Ctesebius, 164n

Dall'Orto, Giovanni, 43
Daston, Lorraine, 33
Dasypodius, 26, 28. *See also* mechanical time; clock, pendular
"dead eye," 5, 46, 49, 158, 159, 163n. *See also* Arsic, Branka; Descartes, Rene

Dear, Peter, 15, 33, 46, 59, 60, 63–64, 169n, 189n–190n; and *Discipline and Experience*, 185n, 186n
Deleuze, Gilles, 19, 55, 180n–181n
Descartes, Rene, 5–8, 68, 76, 154, 158, 170n, 183n, 190n; letter to Vatier, 6; theory of perception of, 5. *See also* Arsic, Branka; "dead eye"
desexualization, 8, 9, 10, 19, 21, 23, 60, 63, 173n, 180n–181n
dildo, 55, 171n. *See also* sex, instrumentation of
Diotima, 41
Dollimore, Jonathan, 56
Donaldson, Arnold, 9
Donohue, John W., 183n–184n, 189n
Drake, Stillman, 126, 130, 164n
Dreyfus, Hubert L., 192n
Dubarle, Dominique, 46
Durkheim, Emile, 150

Eglin, Trent, 197n
elementary machine: invented by Jesuits, 183n
Elias, Norbert, 20–21
Enlightenment, 2, 37, 38, 68
Eros, 41–42, 43, 153
erotic improvisation, 3, 10, 81. *See also* Butler, Judith
ethnomethodology, 11–13, 15, 82, 131, 192n. *See also* Garfinkel, Harold; Livingston, Eric
experiment: 39, 40, 63, 150, 185n; and "objectivity," 48; and virtual witnessing, 46

Feldhay, Rivka, 61
Feyerabend, Paul, 173n
Ficinio, Marsilio: translator of Plato, 43
Findlen, Paula, 35, 44, 45, 174n
Fiorenza, Elizabeth Schussler, 68–69
flagellation, 28, 59, 188n, 189n
Foucault, Michel: 13, 20, 25, 37, 49, 50, 81; on Aristotle, 42; on bodies and pleasure, 157n; and "body-instrument link," 15, 82–86, 192n; on celibacy,

Index

Académie Montmor, 64
Accademia del Cimento, 64, 65
Accademia degli Incogniti, 44
Accademia degli Intronati, 44
Accademia dei Lincei, 7, 45, 65–66, 182n; sexual politics of, 48–49
Adorno, Theodor, 175n
alchemy, 42, 178n, 179n, 196n–197n
Alliez, Eric, 20, 165n–166n
Ambrose, of Milan, Saint, 23
amor Socraticus, 43, 45
Anderson, Benedict, 27. *See also* mechanical time
apathy: and seventeenth-century science. *See* Boyle, Robert; Kant, Immanuel; Sade, Marquis de
Aretino, Pietro, 44, 173n–174n
Aristotle: and sensual/conceptual relation, 41–43. *See also* physics
Aristotle's Master-piece: 17th-century sex manual, 172n–173n
Arsić, Branka, 163n
asceticism, 3, 14, 15, 23, 33, 56, 60–61, 65, 68, 155. *See also* Jesuits
Ashworth, William B., 64, 65, 186n
askēsis, 2, 21, 42, 66
Augustine, Saint, 23, 42, 67

Bacon, Francis, 34, 172n, 173n–174n, 178n–179n
Ballard, J.G., 31. *See also* speed; technophilia

Barkan, Leonard, 45
Barthes, Roland, 73–74, 155, 180n
Basil, Saint, 23
Baudrillard, Jean, 35, 54
Bede, Saint, the Venerable: author of *De temporibus* and *De temporum ratione*, 26. *See also* time
Benedict, Saint, 23
Benedictines: and bodily discipline, 25
Benjamin, Walter, 20, 27. *See also* mechanical time
Bersani, Leo, 55
Biagioli, Mario, 4, 48, 65–66
biophilia, 30
Bloch, Iwan, 27–28.
Bloor, David. 33
body-instrument link, 13–14, 15, 77, 82–83, 87, 93, 113, 150, 158, 192n. *See also* Foucault, Michel; pendulum; prism
Bono, James, 34
Book of the Courtier, 46. *See also* Castiglione, Baldassare
Borsellino, Nino, 44
Boswell, John, 66
Boyle, Robert, 8, 172n; and air pump, 174n; and "apathy," 34, 36–37, 38; and "the gentleman," 33–34, 36, 174n; and "Hook," 11; and scientific experimentation, 196n; and, 174n; and "virtual witnessing," 11, 50. *See also* experiment; seventeenth-century science

CONCLUSION

1. Freud, *Civilization and Its Discontents,* trans. James Strachey (New York & London: W. W. Norton & Company, 1989), 81.

2. Michel Foucault, *Discipline and Punish: The Birth of the Prison,* trans. Alan Sheridan (New York: Vintage Books, 1979), 162.

3. Luce Irigaray, *An Ethics of Sexual Difference,* trans. Carolyn Burke and Gillian C. Gill (Ithaca: Cornell University Press), 5.

4. Judith Butler, *Gender Trouble: Feminism and the Subversion of Identity* (New York & London: Routledge, 1990), 146.

5. Ibid., 148.

6. Foucault, "Nietzsche, Genealogy, and History," in *The Foucault Reader,* ed. Paul Robinon (New York: Pantheon Books, 1984), 148.

7. Foucault, Introduction to Georges Canguilhem, *The Normal and the Pathological,* trans. Carolyn R. Fawcett in collaboration with Robert S. Cohen (New York: Zone Books, 1991), 100.

8. For the context in which Foucault explored techniques of pleasures see Gayle Rubin, "The Catacombs: A Temple of the Butthole," in *Leatherfolk: Radical Sex, People, Politics, and Practice,* ed. Mark Thompson (Boston: Alyson, 1991), 119–41; Geoff Maines, *Urban Aboriginals: A Celebration of Leathersexuality* (San Francisco: Gay Sunshine Press, 1984), *Gentle Warriors* (Stamford, Conn.: Knights Press, 1989), "The View From a Sling," in *Drummer* 121, 31–35; Jack Fritscher, "Upstairs Over A Vacant Lot . . . the Catacombs," in *Drummer* 23, 1978.

9. Michel Serres, *"Ambrosia and Gold,"* in *Calligram: Essay in New Art History from France,* ed. Norman Bryson (Cambridge: Cambridge University Press, 1988), 11, 6, 130.

10. Ibid., 118.

11. Ibid., 117.

knowledge, was at the same time, in Husserl's parlance, a method of concealing its foundations. Trent Eglin makes a distinction between the "esoteric" (implicit) and "exoteric" (explicit) aspect of laboratory work. The first is a self-reflective examination of the laboratory practices while the latter is an explicit and formal structure of what was "esoterically" achieved. Corpuscular and materialistic chemistry, in producing the public protocol for experimentation, pushed alchemists from laboratories. As a consequence of this process, an amnesia developed about the "esoteric" dimension of laboratory work. Now viewed as the idiosyncracy of a radical individualism, "esoteric" foundations of alchemy were replaced by the formal and public protocol of new laboratory sciences. This, however, does not mean that the "esoteric" aspect of experimental work has vanished from laboratories, but only that it is hidden as an implicit presupposition of the explicit structures. See Trent Eglin, "Introduction to a Hermeneutics of the Occult: Alchemy," in *On the Margin of the Visible,* ed. Edward A. Tiryakian (New York: John Wiley & Sons, 1974), 323–50. Accordingly, this paper attempts to discover the "esoteric" structures of Galileo's pendulum.

19. However, the very idea that the order of the world can be revealed by the use of a human-made object was for many scholars of that time a demonic one. Many of them refused to even look through the telescope since their theoretical a priori flatly rejected it as anything meaningful. In fact, it was viewed as rather deceptive (Drake, *Galileo at Work,* 166).

20. Galileo has chosen three lengths of 4:9:16, probably because of the time square rule that he had discovered earlier in his inclined plane experiment (Stillman Drake, "The Role of Music in Galileo's Experiments," *Scientific American* 232 (June 1975): 98–104). The rule claims that the acceleration of a falling body is constant and is proportional to the squares of elapsed time units. If an object is falling through equal time units, first, second, third, fourth, nth, then the acceleration of the falling body through these units increases constantly as the squares of these units, namely, 1, 4, 9, 16, ... nth. The pendulum is an example of the free falling body along the circular line. Since the pendular motion is isochronous, then the time of the swing depends solely upon its lengths (presuming that all friction is eliminated). If the lengths are selected as if squares of time in the free-fall 4" 9" 16" then the motion must conform to the square root of their lengths but in an inverse order, the square root of the longest is the number of swings of the shortest and reverse.

21. I found in a manual of experimental physics the following instructions indicating that my decision to use the terminal plane as the counting measure was correct: "Since it is difficult to determine just when the pendulum reaches the end of its swing, it is better to note the time at which it passes through its equilibrium position" (L. R. Ingersoll and M. J. Martin, *A Laboratory Manual of Experiments in Physics* [New York: McGraw-Hill Book Company, 1942], 63).

22. Martin Heidegger, *Being and Time* (San Francisco: Harper, 1962), 98.

23. Harold Garfinkel and Lawrence D. Wieder, "Two Incommensurable, Asymmetrically Alternate Technologies of Social Analysis," *Text in Context* (London: Sage Publications, 1992), 191.

24. Ibid. 191–92.

9. Livingston, "The Idiosyncratic," 9.

10. Galilei, *Dialogues Concerning,* 107; *Opere,* vol. 8, 150.

11. Shapin and Schaffer detect a special set of techniques required for the replication of scientific instruments and the corresponding phenomena (*Leviathan,* 226). Also, see Simon Schaffer, "Glass works: Newton's prisms and the uses of experiment," in *The Use of Experiment: Studies in the Natural Sciences,* ed. David Gooding, Trevor Pinch, and Simon (Cambridge: Cambridge University Press, 1989); Harry Collins, *Changing Order: Replication and Induction in Scientific Practice* (London: Sage, 1985).

12. Galilei, *Dialogues Concerning,* 107; *Opere,* vol. 8, 150.

13. Also it is interesting to note that the suggested lengths by Galileo are not correct. According to Viviani who actually tested the pendulum, the correct lengths are 64/16, 64/9, 64/4; that is 16, 7 1/9, and 4. See Galileo, *Two New Sciences* (Madison: University of Wisconsin Press, 1974), 107n.

14. In Jacques Derrida, *Edmund Husserl's Origin of Geometry: An Introduction* (Boulder, Colo.: Nicolas Hays Ltd., 1978), 60.

15. David Gooding points out that in order to communicate novel facts through the use of their instruments, scientists must pass on the *know-how* that produces them, including mathematical proofs as well, see David Gooding, "Mathematics and Method in Faraday's Experiment," unpublished paper, 1992: 10–11. Scientific conventions presume that the instructive respecification, science as the *know-how,* is institutionalized as a familiar habit. Shapin and Schaffer mark the time when a discovery becomes a convention in the case of Boyle's air pump: "the moment when skill in making pumps had been transmitted, when a replica of a pump could be said to have been produced, when that replica had produced the same phenomenon as that reported by Boyle, and when a phenomenon could count as a challenge to Boyle's own claims" (*Leviathan,* 226).

16. Husserl, *The Crisis,* 29.

17. Shapin and Schaffer write: "A fact is a constitutively social category: it is an item of public knowledge. We displayed the process by which a private sensory experience is transformed into a publicly witnessed and agreed fact of nature. In this way, the notion of replication is basic to fact-production in experimental science" (*Leviathan,* 225).

18. The order of the witnessing region in Galileo's instruments was later canonized into a scientific convention for generating facts. Early experimentalists, like Boyle, insisted on this model of collective witnessing as a reaction towards the Alchemists' idiosyncratic and radical individualism of experimentation (Shapin and Schaffer, *Leviathan,* 55–60). Boyle used the justice system of multiplication of evidence and a collective jury as a model for a public verification of scientific experiments. He questioned whether Alchemical experiments could ever be replicated since they were only thought experiments and were never tried. The replication of the claimed phenomenon by anybody is the key to scientific objectivity. Shapin and Schaffer write: "Many phenomena, and particularly those alleged by alchemists, were difficult to accept by those adhering to the corpuscular and mechanical philosophers. Boyle averred that they have seen them can much more reasonably believe them, than they that have not" (*Leviathan,* 56).

The model of science that came out of Galileo's instruments is a model which, while becoming an explicit and procedural method for generating objective

25. MacLachlan, "Galileo's Experiments."
26. Ariotti, "Galileo on Isochrony."
27. Ibid., 420.
28. MacLachlan, "Galileo's Experiments."
29. Drake, *Galileo at Work*, 66–68.
30. Galilei, *Dialogues Concerning*, 178; *Opere*, Vol. 8, Proposition 6. Theorem 6, 221.
31. Drake, *Galileo at Work*, 71.
32. Galilei, *Dialogues Concerning*, 211; *Opere*, Vol. 8, Proposition 36. Theorem 22, 262.
33. Drake, *Galileo at Work*, 71.
34. Ariotti, "Galileo on Isochrony," 423.
35. Ibid., 424.

CHAPTER 6. THE RESPECIFICATION OF GALILEO'S PENDULUM

1. Drake and MacLachlan, "Galileo's Discovery," 109.
2. Thomas Settle, "An Experiment in the History of Science," *Science* 133: 21.
3. Eric Livingston, "The idiosyncratic specificity of the methods of physical experimentation," *Australian and New Zealand Journal of Sociology* 31, no. 3 (November 1995): 7.

According to Livingston, Galileo has invented our "Received View" from which we, when the swinging pendulum demonstrates the causal relation between the length of a pendulum's cord and its period, *see* motion in terms of abstract mathematical claims rather than what is in front of our eyes that is, as Livingston points out, "an organization of practices that exhibits the adequacy" (Livingston, 11) of the described relation between the pendulum length and period. As Livingston writes:

> Physicists, both theoretical and experimental, realize the essential idealization involved in the Received View. In practice, a 'domain of worldly phenomena' is always understood as a domain of experimentation. Moreover, every term, operation, and law in the propositional space is understood as a reference to experimental settings and to a situation of inquiry into those settings. (Livingston, 7)

And yet, in spite of the physicist's practical reference to Galileo's science, scientific scholarly books ignore the obvious disparities between the production and representation of science and continue to represent the intricacies of time, length and gravity of Galileo's pendulum only as an abstract formula

$$T = 2\pi \sqrt{l/g}$$

4. Drake, *Galileo at Work*, 69.
5. Drake, *Galileo: Pioneer Scientist* (Toronto: University of Toronto Press, 1990), 16, 19.
6. Ibid. See also Hill, "Pendulums."
7. Galilei, *Dialogues Concerning*, 97; *Opere*, vol. 8, 139.
8. Ibid., 97–98.

8. Edmund Husserl, *The Crisis of European Sciences and Transcendental Phenomenology* (Evanston, Ill.: Northwestern University Press, 1970), 27, 32, 34, 36, 49.

9. See Drake, "Galileo's Discovery," 84–92.

10. Ibid. See also Koyré, *Galileo's*, 95–109.

11. Dominique Dubarle, "Galileo's Methodology of Natural Sciences," in *Galileo Man of Science,* ed. Ernan McMullin (Princeton, N.J.: The Scholar's Bookshelf, 1988), 305–8.

12. See Robert Naylor, "Galileo's Experiments with Pendulums: Real and Imaginary," *Annals of Science,* 33, 1976, 106.

13. In one of his writings, Galileo expressed pedagogical standards which illustrate what might have been the real purpose of the pendulum:

> For just as there is no middle ground between truth and falsity in physical things, so in rigorous proof one must either establish this point beyond any doubt or else beg the question excusable, and there is no chance of keeping on one's feet by invoking limitations, distinctions, verbal distortions, or other fireworks; one must with but few words and at the first assault become Caesar or nobody. (Quoted in Naylor, "Galileo," 134)

14. About the relation between the pendulum's geometric structure and the geometrization of scientific practices see the next chapter.

15. The pendulum's motion appears to the human eye as isochronic but it is not mathematically so. Only certain, though unobservable curves of the pendulum's arc are isochronic. (Dutch mathematician Christian Huygens became the first to determine the isochronic curve of the pendular arc. See Christian Huygens, *The Pendulum Clock or Geometrical Demonstrations Concerning the Motion of Pendula as Applied to Clock* (Ames: Iowa State University Press, 1986), 69–71. It is interesting to observe that Galileo rests his theorem of the pendulum's isochronicity on the experience as much as on the mathematical proof. That Galileo was aware of this see David K. Hill, "Pendulums and Planes: What Galileo Didn't Publish" *Nuncius* 9 (1994): 499–515.

16. Piero Ariotti, "Galileo on the Isochrony of the Pendulum," *Isis,* vol. 59, 1968, 414Ff.

17. James MacLachlan, "Galileo's Experiments with Pendulums: Real and Imaginary," *Annals of Science,* no. 33, 1976, 174.

18. Ariotti, "Galileo on Isochrony," 416.

19. Ibid.

20. Koyré, *Metaphysics,* 1.

21. Stillman Drake, *Galileo at Work* (Chicago: University of Chicago Press, 1981), 65.

22. The following pendula experiment is given in Galileo Galilei, *Dialogues Concerning the Two Great Systems of the World,* trans. Thomas Salusbury (London: 1661), 97.

23. Piero Ariotti, "Aspects of the Conception and Development of the Pendulum," *Archive for History of Exact Sciences* 8 (1972), 355.

24. Ariotti, "Galileo on Isochrony"; MacLachlan, "Galileo's Discovery"; David K. Hill, "Pendulums and Planes: What Galileo Didn't Publish," *Nuncius, IX,* 1994, 499–515.

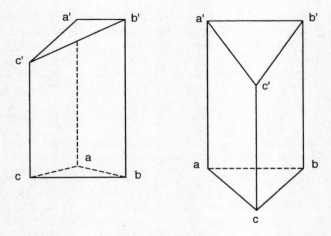

Figure 39

16. Garfinkel, H. et al, "Respecifying the natural sciences as discovering sciences of practical action, I and II: Doing so ethnographically by administering a schedule of contingencies in discussion with laboratory scientists and by hanging around their laboratories" (unpublished manuscript, 1988).

17. Edmund Husserl, *The Crisis of European Sciences and Transcendental Phenomenology* (Evanston, Ill.: Northwestern University Press, 1970), 43; also see Aaron Gurwitsch, *Phenomenology and the Theory of Science* (Evanston, Ill.: Northwestern University Press, 1974); "Galilean Physics in the Light of Husserl's Phenomenology," in *Galileo Man of Science,* ed. F. McMullin (Princeton, N.J.: The Scholar's Bookshelf, 1988).

18. Husserl, *The Crisis,* 52.

CHAPTER 5. THE FORMAL STRUCTURE OF GALILEO'S PENDULUM

1. Thomas S. Kuhn, *The Structure of Scientific Revolution* (Chicago: University of Chicago Press, 1970).
2. Marshal Clagett, *The Science of Mechanics in the Middle Ages* (Madison, Wis.: University of Wisconsin Press, 1961), 536–71.
3. Alexandre Koyré, *Galileo's Studies* (New Jersey: Humanities Press, 1978), 73.
4. Ibid., 109.
5. See Stillman Drake, *Discoveries and Opinions of Galileo* (Garden City: Doubleday & Co. Inc., 1967), 237–38.
6. Drake, "Galileo's Discovery of the Law of Free Fall," *Scientific American* 228, no. 5 (May 1973): 84–92
7. Alexandre Koyré, *Metaphysics and Measurement: Essays in Scientific Revolution* (Cambridge, Mass.: Harvard University Press, 1968), 37.

2. For my part to avoid this temptation, I rely on Foucault's history of sexuality and on ethnomethodology. From Foucault I adopt the view that not any body but the historian's body ought to be vindicated from the disciplinary discourse and installed as the history's referent. I would, however, have betrayed this resistance had I stayed only with rewriting Foucault's history. Employing ethnomethodology allowed me to use my own body as a *method* for rescuing Galileo's pendulum from being a mere *sign* of history to being a *practice,* as it was for Galileo. While ethnomethodology will allow me to insert my *method* into the text, Foucault's history of sexuality will historically contextualize this method. Only by merging these two asymmetric methodologies could I find a satisfactory ground to start an inquiry into the methodology of the "body-instrument link," which is my real focus here.

3. Michel Foucault, *Discipline and Punish: The Birth of the Prison,* trans. Alan Sheridan (New York: Vintage Books, 1979), 152–53.

4. See David M. Halperin, *Saint Foucault: Towards a Gay Hagiography* (Oxford: Oxford University Press, 1995).

5. Foucault, *Discipline,* 152–53.

6. Ibid.

7. Ibid.

8. See Foucault, *Archeology of Knowledge* (New York: Pantheon Books, 1972), 7. Hubert L. Dreyfus and Paul Rabinow explain Foucault's decontextualization: "Studying discursive formations requires a double reduction. Not only must the investigator bracket the *truth* claims of the serious speech acts he is investigating—Husserl's phenomenological reduction—he must also bracket the meaning claims of the speech acts he studies; that is, he not only must remain neutral as to whether what a statement asserts as true is in fact true, he must remain neutral as to whether each specific truth claim even makes sense, and more generally, whether the notion of a context-free truth is coherent"(Michel Foucault, *Beyond Structuralism and Hermeneutics* [Chicago: University of Chicago Press, 1983], 49).

9. Judith Butler, *Bodies that Matter: On the Discursive Limits of "Sex"* (New York: Routledge, 1993), 12–16.

10. Alfred Schutz, "The Problem of Rationality in the Social World," in *Collected Papers,* vol. 2 (The Hague: Martinus Nijhoff, 1973); Harold Garfinkel, *Studies in Ethnomethodology* (Englewood Cliffs, N. J.: Prentice-Hall, 1967).

11. Foucault, *The Use of Pleasure: The History of Sexuality,* vol. 2, trans. Robert Hurley (New York: Vintage Books, 1990), 26–27.

12. Ibid., 13.

13. This text is made intelligible through the use of a single prism. A particular type of prism must be used in order to understand instructions in this text. Its specifications are pictured in figure 39. The reader can inquire about information on where to buy this kind of prism by writing to: Sales Assistance: Bob Cross—TEDCO, INC. 4985. Washington Street, Hagerstown, IN 47346-1596, 1-800-654-6357; in Indiana call 1-317-489-4527, Fax 317-489-5752. The catalog item number of this prism is # 00010. Also, http://www.tedcotoys.com/contact.cfm.

14. Bjelić, D. and Lynch, M. (1993), "Goethe's 'Protestant Reformation' as a textual demonstration," *Social Studies of Science* 4: 703–24.

15. Michael Polanyi, *The Tacit Dimension* (Garden City, N.Y.: A Doubleday Anchor Book, 1967), 18.

"monastic perfection" and "harsh discipline" (Noble, *The World,* 218–19). In both the monastic and the experimental settings, activities are controlled according to a "timetable": actions are broken down into their simplest elements to which direction and duration were assigned and prescribed. Experiments have to operate as a "protected place of disciplinary monotony" (Foucault, *Discipline,* 141) in which individuals are related functionally to each other and through a higher rationality. Adopting the cell's operation as a "unit" of network relations and ranks, the experiment creates its "analytical space" (ibid., 143). Discipline in the laboratory, as in the cell, defines the body's relation to the objects of manipulation; the discipline decomposes the conduct of the natural body and arranges a positive economy of movements, maximizing speed and efficiency. According to Foucault, "A disciplined body is the prerequisite of an efficient gesture" (ibid., 152) which is to say that constructing and controlling the experimental situation, bodies, and instruments, demanded new conditions of rational efficiency. As a new formative scheme of knowledge, the experiment, I argue, came about as a new "productive link" between the body and the instrument (*Discipline,* 153).

60. Barthes, *Sade, Fourier, Loyola,* trans. Richard Miller (New York: Hill and Wang, 1976), 57.

61. Ibid., 44.

62. Ibid., 59.

63. Ibid., 59–60.

64. Ibid., 44.

65. "I think," Kristeva writes, "that for a women, generally speaking, the loss of identity in jouissance demands of her that she experience the phallus that she simply is; but this phallus must immediately be established somewhere; in narcissism, for instance, in children, in a denial and/or hypostasis of the other woman, in narrow minded mastery, or in fetishism of one's "work." See Julia Kristeva, *Desire in Language: A Semiotic Approach to Literature and Art,* ed. Leon S. Roudiez (New York: Columbia University Press, 1980), 164.

66. Judith Butler, *Subjects of Desire: Hegelian Reflections in Twentieth-Century France* (New York: Columbia University Press, 1999), 233.

67. Barthes, *Sade,* 61.

68. This French word may signify joy as well as erotic pleasure.

69. See Leon S. Roudiez, introduction to Kristeva, *Desire in Language,* 16.

70. Koyré, *Metaphysics,* 105–6.

71. Eve Kosofsky Sedgwick, *Between Men, English Literature and Male Homosocial Desire* (New York: Columbia University Press, 1985), 1–5; also, *Epistemology of the Closet* (Berkeley: University of California Press, 1990), 87–88.

72. Galileo, *Dialogue* (G. OP. VII, 170).

73. Alexander Waugh, *Time, Its Origin, Its Enigma, Its History* (New York: Carroll & Graf Publishers, 2001), 20.

CHAPTER 4. THE "BODY-INSTRUMENT LINK" AND THE PRISM: A CASE STUDY

1. See Paul Rabinow (ed.), *Foucault's Reader* (New York: Pantheon Books, 1984), 92.

76n.) thus, since I did not consult the original text and am relying exclusively on Koyré's account, the passages that I am quoting from Koyré might be the problematic ones.

49. Father Riccioli, in his arduous effort to determine a second of time, acknowledged the discipline involved: "This pushing of the pendulum is by no means easy and implies a long training" (Koyré, *Metaphysics,* 104).

50. Ibid.

51. Ibid., 104–5.

52. Ibid., 105–6.

53. Ibid., 107–8.

54. Porter, *The Educational,* 7.

55. Foucault, *Discipline and Punish: The Birth of the Prison,* trans. Alan Sheridan (New York: Vintage Books, 1979), 137.

56. Ibid., 151.

57. Ronald Naylor, who tested the accuracy of Galileo's experimental measurement, speaks to the claim that it depends on the skills and bodily discipline required by the experimental situation. He writes: "Owing to the large volume of the tank there is really no inaccuracy due to loss of head, as suggested by Koyré, even for rates of flow as large as four grams per second. An obvious procedure to adopt is to return the outflow to the tank periodically and so ensure the level in the tank remains nearly constant. In my experience, the major limitation on the accuracy of the measurement appears to lie in the speed of *reaction* [italics mine] of the experimenter" (Ronald H. Naylor, "Galileo's Simple Pendulum," *Physis,* vol. 16, 1974, 29). "The speed of reaction of the experimenter" ensures the accuracy of measurement.

58. The reasons were to improve the Church calendar and commerce by improving the timing of navigation (Jacques Le Goff, *Time, Work and Culture in the Middle Ages* [Chicago: University of Chicago Press, 1982], 35); and to use precise timing as a weapon for colonizing pagan cultures. Jesuits were aware that whoever controls time and can provide precise measurement has power over pagan cultures. How much power they had accrued through their precise measurement of time has been exemplified by the fact that the Jesuits were the only missionary order permitted in China due to their ability to improve the Emperor's calendar and mechanical measurement, which the huge bureaucratic empire heavily depended on (see for this reference, David S. Landes, *Revolution in Time: Clocks and the Making of the Modern World,* [Cambridge: Harvard University Press, 1983], 38; Delia Pasquale M. S. J., *Galileo in China: Relations through the Roman College between Galileo and the Jesuits Scientist-Missionaries [1610–1640],* trans. Rufus Suter and Matthew Sciascia [Cambridge: Harvard University Press, 1960], 4–6).

59. The grounding of "time" in the history of discipline, rather than just in the history of ideas, points to the monastic cell as a disciplinary setting and as a methodic resource for the nascent experimental science. The parallel between the monastic and experimental, I should stress following Foucault, is this. The structure of experimental design usually requires "enclosure," as did the "monastic cell." Marin Mersenne, one of the first seventeenth-century experimentalists, a member of the religious order of Minims, a priest, a brother and an abbot of the Minim convent in Paris, a friend and a teacher of Descartes, was at the center of experimental science in France. His experiments were done in the environment of the monastic cell; his order required in addition to chastity a total abstinence from meat and dairy products, perpetual Lenten fasting,

the pain for the sake of gaining knowledge and laments about the loss of a pain in the ascending modern education: "As to the methods, ever easier and easier, which are being excogitated, whatever convenience may be found in them, there is this grave inconvenience; first, that what is acquired without labor adheres but lightly to the mind" (Thomas Hughes, *Loyola and the Educational System of the Jesuits* [New York: Charles Scribner's Sons, 1892], 291). Donohue makes a similar observation about the pairing of austerity and pleasure in gaining knowledge: "Indeed, more austere spirits have frowned upon this Jesuit interest in providing tangible rewards beyond the satisfactions (often invisible to students) of knowledge itself. . . . It will be enough to note here that a great deal of care was taken to make learning *pleasant*" (John W. Donohue, S.J., *Jesuit Education* [New York: Fordham University Press, 1963], 153). It has been the Jesuit custom to combine the austerity of learning with the various forms of pleasures evident in the very decision to build schools in very pleasing places, to replace corporal punishment by winning "students by love rather than fear" (William J. McGucken, *The Jesuits and Education* [New York: The Bruce Publishing Company, 1932], 33), to substitute prolonged and tiresome study practiced by Benedictines with learning through games, prize-giving, ribbons, insignia and even with playing sports, ballet and performing theater (Gaukroger, *Descartes*, 42–43). Children were taught in the Jesuit schools from the very beginning how to extract the pleasure of reason out of their bodies, and more importantly, that pain and knowledge are coextensive.

39. Loyola, *The Spiritual Exercises of St. Ignatius Loyola*, trans. Elizabeth Meier Tetlow (Lanham, Md.: University Press of America, 1987), 105, 140.

40. Loyola, *The Spiritual*, 108; 141.

41. Elizabeth Schussler Fiorenza, *In Memory of Her* (New York: Crossroad, 1983), 44.

42. Noble, *The World*, xiii.

43. Judith Butler, *Gender Trouble: Feminism and the Subversion of Identity* (New York: Routledge, 1990).

44. Luce Irigaray, *An Ethics of Sexual Difference*, trans. Carolyn Burke and Gillian C. Gill (Ithaca: Cornell University Press), 5.

45. Butler, *Gender*, 9; see also Butler, *Bodies that Matter: On the Discursive Limits of "Sex"* (New York: Routledge, 1993), 8.

46. Dear, *Discipline*, 79.

47. Peter Dear observes that in the seventeenth century natural philosophy theorized not only that the natural universe is mechanical but also that the rules of mechanics constrain and repress the human body. For example, many gentlemanly practices, such as fencing, horsemanship, or dancing are governed by geometrical forms, thus bodies in performing them act as automata. (Dear, "A Mechanical Microcosm," in *Science Incarnate: Historical Embodiments of Natural Knowledge*, ed. Christopher Lawrence and Steven Shapin, [Chicago: University of Chicago Press, 1998], 53). Thomas Settle's respecification of Galileo's inclined plane experiment contains numerous references to the experimenter's repetitive operations, which suggests that the seventeenth-century body of the experimenter may have been subject to the mechanized settings of conduct and act as an automaton too.

48. Dear warns that "Koyré's account is not always reliable, and sometimes combines different passages in questionable ways" (Dear, "A Mechanical Microcosm,"

Cassian, to which he confines the whole of his analysis, leaving aside the question of physical sex. His theme is *immunditia,* something which catches the mind, waking or sleeping, off its guard and can lead to pollution, without any contact with another; and the *libido,* which develops in the dark corners of the mind. In this connection Cassian reminds us that *libido* has the same origin as *libet* (it pleases). (Foucault, 191)

Mind is conditioned to be impure and corrupted if the body and the will are not disassociated from the world. Cassian's "second fornication" pertains to the mind. This is a bodiless "fornication" in which pleasure permeates the mind and makes itself a source of pleasure. Chastity, thus, is a discipline pertaining to sexuality of the mind as much as of the body.

35. Boswell, *Same-Sex,* 157.
36. Porter, *The Educational,* 7.
37. See Stephen Gaukroger, *Descartes: An Intellectual Biography* (Oxford: Clarendon Press, 1995), 39.
38. Flagellation, for example, although a rare case of the bodily discipline among the Jesuits, is a poignant case of the denunciation of "feminine" weaknesses and affirmation of "masculine" endurance. Domination becomes pleasure in itself; the pleasure of penitence becomes a pleasure of moral conduct. Abbe Boileau elaborates this transformation of pain into pleasure in his *Histoire des flagellan . . .* absolutely unavoidable that when the loin muscles are whipped with sticks or with a whip the animal spirits are violently forced back toward the pubic bone and that they excite immodest movements because of their proximity to the genital parts: these impressions go to the brain and paint there vivid images of forbidden pleasures which fascinate the mind and with their tempting charms reduce chastity to dissipation" (quoted in Ann W. Ramsey, "Flagellation and the French Counter-Reformation: Asceticism, Social Discipline, and the Evolution of a Penitential Culture," in *Asceticism,* 584n.) Here, the common-sense binary between the erotic and the painful are blurred. Moreover, because painful discipline is perceived as being a form of masculine self-empowerment, pain also becomes a means of strengthening pleasurable homosocial and even homoerotic ties. Similarly, as we shall see shortly, the Jesuit production of scientific knowledge operates within the same economy, whereby the pain of self-discipline is converted into the pleasure of rationality. The Jesuit education has been another example of the economy of pain and pleasure. The elimination of the dichotomy between the pain associated with various forms of bodily austerity, a pleasure of knowledge and mastery characterizes the Jesuit education. Porter observes, "Most frightful is the truth which is uttered of this society by one of its latest historians, that 'it developed human devotedness to its extremist capacity, and made of the most absolute obedience, a lever, the incessant and ever present activity of which must necessarily take the place of every other species of power'" (Porter, *The Educational,* 8). The authority in Jesuit upbringing, Porter claims, is made absolute by means of austerity. But, as he correctly concludes, "These austerities were no end in themselves, for it was never Loyola's design to train a company of painful ascetics, the only products of whose energy should be bloody flagellations, marvelous fasting, and unnatural self-tortures" (ibid). Father Roothaan, General of the Society in his encyclical letter to the Order justifies

lough and James Brundage (Amherst, N.Y.: Prometheus Books, 1994), 22–33; Charles A. Frazee, "The Origins of Clerical Celibacy in the Western Church," Church *History* 41, no. 2 (June, 1972): 149–67; Peter Brown, *The Body and Society: Men, Women, and Sexual Renunciation in Early Christianity* (New York: Columbia University Press, 1988); Bruce S. Thorton, *Eros: The Myth of Ancient Greek Sexuality* (Boulder, Colo.: Westview Press, 1997); John Boswell, *Christianity, Social Tolerance, and Homosexuality: Gay People in Western Europe from the Beginning of the Christian Era to the Fourteenth Century* (Chicago: The University of Chicago Press, 1980); Boswell, *Same-Sex Unions in Premodern Europe* (New York: Villard Books, 1994); Noble, *The World;* Max Weber, *Economy and Society,* vol. I (Berkeley: California University Press, 1978); Wimbush and Valantasis (Eds.), *Asceticism.*

30. Frazee, "The Origins," 167.
31. Boswell, *Same-Sex,* 159.
32. Augustine, *Confessions,* 48.13.
33. Boswell, *Same-Sex,* 161.
34. The rational structures of an experimental conduct in this respect is an outcome of the monk's relation to the body and pleasure. The Christian "fear of fornication" and the code of celibacy, Foucault posits, was at the core of this relation. Historically speaking celibacy started as a voluntary sexual practice among the few radical Christians only to became a law for all the clergy in the sixteenth century. In the twelfth century Cassian discusses chastity in the six chapters of his *Institutiones,* arriving at the idea that the whole essence of the fight for chastity is not to target what is between two sexually related individuals, but rather to target a different reality. Cassian's instructions ignore physical sex and introduce chastity as a self-formative and self-empowering discipline. Abstaining from sex is not a passive but an active attitude towards the body, pleasure and transformative of an identity. Cassian offers six signs of the identity transformation as a result of achieving chastity. First, the monk awakes without being "smitten by a carnal impulse"—*impugnatione carnal non eliditur;* second, not allowing "voluptuous thoughts"—*voluprariae cogitationes*—to dwell on the monk; third, when one can look at women without any feeling or desire; fourth, when one no longer feels any movement of flesh during one's waking hour; fifth, when reading, discussing or contemplating sexual content has no effect on the monk; and sixth, when a seductive women does not emerge in one's dream. These signs or stages of transformation aim at the monk's "disinvolvment of the will." The first step here, according to Foucault, is to exclude the will from bodily involvement in the world; second, to exclude the will from imagination; third, to exclude the will from the action of the senses; fourth, to exclude the will from figurative thinking; and, finally, from oneiric involvement (Foucault, "The Battle for Chastity," in *Ethics: Subjectivity and Truth,* ed. Paul Rabinow, trans. Robert Hurley et al. [New York: The New Press, 1997], 190).

Chastity as a method of rational conduct and a technology of the self stresses the relation between reason and pleasure. Foucault elaborates this relation:

> [T]he bodily and mental reflexes that bypass the mind and, becoming infected with impurity, may proceed to corruption, and on the other side an internal play of thoughts. Here we find the two kinds of "fornication" as broadly defined by

22. Dear, *Discipline and Experience*, 55.

23. Ibid. 75.

24. Alexandert Koyré, *Metaphysics and Measurement: Essays in Scientific Revolution* (Cambridge: Harvard University Press, 1968), 102–8; Peter Dear, *Discipline and Experience*, 76–79.

25. Collaborative character of the Jesuit science is partly due to their self-inflicted reform of their education and sciences. Because their Ptolemaic science was in principle incorrect, it needed constant corrections, modifications, and alterations in accordance with the reports coming from navigators and missionary astronomers about their newest observations. These reports would be recorded and turned into new educational material. One such reforming enterprise was done by Fathers Riccioli and Grimaldi in the book *Geographia Reformata*. Their goal was to rewrite geography by means of astronomy. Jesuit missionary astronomers were sending their reports from various parts of the earth into one gathering center. The Reverend Hughes writes:

> The first eclipse of the moon which he [Riccioli] makes mention of, among his astronomical reports, had been observed on the night of November 8, 1612, by Father Scheiner at Ingolstadt and by Father Charles Spinola at Nagasaki in Japan. At the time that Riccioli was writing the Jesuit missionaries had multiplied in China. Adam Schall died in 1666, holding the post of President of the Mathematical Tribunal at Peking; he was followed by Ferdinand Verbiest; and then a long line of imperial astronomers of the Celestial Empire, Koeglere, Hallerstein, Siexsas, Frabcesco, De Roche, Espinh, continued to send their reports, either to the colleges of their respective Provinces, or to other mathematical centers, or to the learned societies in Europe, whereof not a few Jesuits were members. Meanwhile, scientific returns from Hindustan, Siam, Tibet, on one side of the globe, and from San Domingo on the other side, poured into the College Louis-le-Grand, and made of this educational center an indispensable auxiliary to the Bureau of Longitudes. (Hughes, *Loyola*, 169–70)

26. Ashworth, "Catholicism," 155.

27. Dear, *Discipline and Experience;* see also Dear, "From Truth to Disinterestedness in the Seventeenth Century," *Social Studies of Science* 22 (1992): 619–31. But their science was not without its shortcomings. It was eclectic and not sufficiently discriminatory in regard to their focus of study (Ashworth, "Catholicism," 155). This eclecticism, according to Ashworth, was due to "the lack of any philosophical superstructure holding the facts being presented" (ibid., 156) complicated by the lack of disciplinary focus because of their eclecticism, their emblematic rather than analytical view of nature, their lack of analytical focus, their probabilism and relativism in explanation. Because of their emblematic approach to nature and their probabilism, for Jesuits, Copernicanism and Galileo's science should be discussed only as hypotheses and plausible explanations, rather than as an explanation of the actual order of things—the view against which Galileo pressed hard to bring to the surface the inconsistency and hypocrisy of their position.

28. Biagioli, "Homosociality," 158.

29. About the history of chastity see Jo Ann McNamara, "Chaste Marriage and Clerical Celibacy," in *Sexual Practices and The Medieval Church,* ed. Vern L. Bul-

sensibility in the court and skillfully used the court power to legitimize his discoveries, he clearly demonstrated a lack of professional tact in relation to the Jesuits. As Father Grienberger, a holder of the chart in astronomy at the Collegio Romano, commented in 1634, "If Galileo had known how to retain the affection of the fathers of this College, he would have lived gloriously before the world and none of his misfortunes would have happened, and he would have been able to write as he chose about everything, even the motion of the earth . . ." (Langford, *Galileo*, 195–96).

12. Ibid, 240–55. About the trial from the standpoint of the Church see pages 137–58.

13. See Rivka Feldhay and Michael Heyd, "The Discourse of Pious Science," *Science in Context* 3, 1989: 109–42. For the differences between Jesuit and Protestant asceticism see Noah Porter, *The Educational Systems of the Puritans and Jesuits Compared*.

14. Feldhay, *Galileo and the Church*, 80, 123. See also Ramsey, "Flagellation," 576.

15. Dear states,

> The "experiment" as a hallmark of modern experimental science, then, is constituted linguistically as a historical account of a specific event that acts as a warrant for the truth of a universal knowledge-claim. "Experiments" in this sense only became a part of a coordinated knowledge-enterprise during the course of the seventeenth century. Understanding how the change came about, and discovering its philosophical meaning, amounts to investigating the cognitive and disciplinary categories that constrained and allowed it. Doing this then draws attention to another crucial difference between scholastic "experience" and modern "experiments" as warrants for statements about nature: the former could only be observational perceptions of nature's ordinary course, whereas the latter by design subverted nature. (Peter Dear, *Discipline and Experience*, 6)

The Jesuits, according to Dear, transformed the nature of their discourse in three distinct ways: a) by transforming the meaning of scholastic "experience" from a series of direct observations to an experimental design based on a mathematical universal knowledge about the natural world that is "justified on the basis of singular items of individual experience" (Dear, *Discipline and Experience*, 13); b) by transforming discourse on mathematics into "a nonlocal philosophical culture," and c) by transforming "mathematical-physics" and "experiment" into a new "experience" that emerged in seventeenth-century science as the vanguard of a new natural philosophy with ambitions to "measure all things and by becoming a science to grasp everything" (Ibid.).

16. Ibid., 12.

17. W. E. Knowles Middleton, *The Experimenters: A Study of the Accademia del Cimento* (Baltimore and London: Johns Hopkins University Press, 1971), 1.

18. Heilbron, *Electricity*, 2.

19. Dear, *Discipline and Experience*, 9, and William B. Ashworth, Jr., "Catholicism and Early Modern Science," in *God and Nature*, ed. David Lindberg and Roland L. Numbers (Berkeley: University of California Press, 1986), 154.

20. Ashworth, "Catholicism."

21. Middleton, *The Experimenters*, 4.

The aim of Jesuit education that was to permeate every aspect of the curriculum were "enlargements to the personality by genuine competitive additions—growth through the acquisition of new habits or skills which represented a qualitative enrichment of intelligence. Information and formation . . . are not mutually exclusive but are rather points along the continuum of learning and one does not exist without the other" (Donohue, *Jesuit Education,* 148–49). Acquisition of skills forms Jesuit students, but this acquisition could not occur without self-instructive learning, emphasizing that understanding is a procedural and practical achievement. "Habits are only developed by practice and that practice is most fruitful when the practitioner understands what he's doing. A multitude of small-scale investigations have shown that pupils who are taught how to study or given the rationale of a procedure learn more successfully than those who have not received such instruction. The Jesuit pedagogy was thoroughly committed to both these principles and aimed first at securing understanding and then at developing mastery" (Donohue, *Jesuit Education,* 149).

9. Ibid., 240–55.

10. The Jesuit strategy of the open dialogue in order to control the opponent was articulated perhaps most eloquently by Father Roothaan, the general of the society, a few centuries back when he addressed the order:

> In Physics and Mathematics we must not prove false to the traditions of the Society, by neglecting these courses which have now mounted to a rank of the highest honor. If many have abused these sciences to the detriment of religion, we should be so much the farther from relinquishing them on that account. Rather, on that account, should the members of the Order apply themselves with the more ardor to these pursuits and snatch the weapons from the hands of the foe, and with the same arms, which they abuse to attack the truth, come forward in its defence. For truth is always consistent with itself, and in all the sciences it stands erect, ever one and the same; nor is it possible that what is true in Physics and Mathematics should contradict truth of a higher order. (Quoted in Hughes, *Loyola,* 292)

11. Feldhay, *Galileo and the Church,* 247–49. Jerome J. Langford describes the Galileo-Jesuit dialogue as complicated and of old standing:

> Another area that has been studied well in the last two decades is the complex relationship that developed between Galileo and the Jesuits. Galileo's long-standing relationship with the Jesuits dates back to his study of lecture notes from the leading professors at the Collegio Romano and correspondence with the leading Jesuit astronomer Christopher Clavius in 1588, through the confirmation by Jesuit astronomers of his early telescopic discoveries in 1610, all the way to his regarding them as allies when Dominicans carried the scriptural issue to the fore in 1615. (Jerome J. Langford, *Galileo, Science and the Church* [Ann Arbor: University of Michigan Press, 1971], 195–96)

The antagonism between Galileo and the Jesuits rose after Galileo's publication of *Il Saggiatore,* Langford observes. Galileo showed impatience with the Jesuits' anti-Copernicanism, and became more aggressive, witty, and tended to ridicule them, thereby eliciting bad feelings and revenge. If Galileo showed incredible professional

The key to Jesuit pedagogical success is twofold: the first is that of employing bodily discipline as a basis for learning. Sister M. Augustine Sheele takes issue in her writings about the educational aspects of Loyola's spiritual writings, with the lack of active learning in modern pedagogy, simultaneously stressing the pragmatic method of Jesuit education that preceded John Dewey's principles of education. She emphasizes that the spiritual pragmatism of "insistence upon domination of man's conduct by reason" in Jesuit education stems from the Jesuits' fundamentally pragmatic attitude towards religious education (Sheele, *Educational Aspects of Spiritual Writings* [Milwaukee, Wisc.: St. Joseph Press, 1940], 81). Pragmatic action was central to Loyola's educational teachings. The Reverend Thomas Hughes succinctly sums up the pragmatism of Jesuit education as "the best way best adjusted to circumstances" (*Loyola*, 142). In the words of Donohue, Loyola's "own ascetical doctrine firmly emphasized the necessity of overcoming indolent passivity and that accent on action becomes a central pedagogical principle both in the *Constitutions* and in the *Ratio*. The reason is clear in every case: if you will the end, you must will the means. But no growth is possible either in holiness or in secular wisdom without enormous effort. The goal which is authentic learning is acquired only by a learner's self-activity" (John W. Donohue, *Jesuit Education* [New York: Fordham University Press, 1963], 148). The reader can also find a reference to Jesuit educational pragmatism and its similarity with James, Dewey, and Mead, in Donohue, *Jesuit Education*, 23. The second key to Jesuit pedagogical success is their teaching through instruction and instrumental demonstrations. Jesuit schools were known to have the best instruments among all the academies; instruments for astronomy, geodesy, drilling, drawing, and sometimes physics. Heilbron emphasizes the importance of textual and instrumental relations in Jesuit education: "Descriptions of these instruments, some built to original design, was a staple in the technical books that the Jesuits produced in profusion in the seventeenth century" (Heilbron, *Electricity*). They regularly updated their stocks of instruments and regularly were "asked for advice about the procurement of instruments, and sometimes acted as intermediaries between purchasers and makers" (Heilbron, *Electricity*, 103–4). They themselves published texts about how to make and use instruments, such as Schott on the pantograph and Zucchi on the elementary machine, both of which were widely circulated. Their literature served as a source of valuable data even to those who were not friends of the Order. Huygens, for example, who was otherwise hostile to the Jesuits, wrote in a letter to Descartes: "For these scribbles can serve you in matters *quae facti sunt, non juris*. They have more leisure than you to provide themselves with experiments" (Heilbron, *Electricity*, 106).

Classroom hierarchy and the learning of science were intertwined in Jesuit education. Classrooms were run on a military model. All students were divided into groups of ten, called *decuriae,* and every group had a head known as a *decurio*. Serving as both student and supervisor gave these students a special pedagogical opportunity to learn through instructing others. This same method has proved successful in modern science. In 1961, the *New York Times* published an article about a puzzle facing the graduate faculty of physics at Harvard: "the number of outstanding students it received from an otherwise undistinguished small college." Inquiry showed that this college had a physics professor who let his seniors conduct the freshmen classes. They learned a good deal of science in the process. In the Jesuit schools, however, necessity was the first inspiration for this monitorial practice" (Donohue, *Jesuit Education*, 66).

HowExpert Guide to Fountain Pens

101+ Lessons to Learn How to Find, Use, Clean, Maintain, and Love Fountain Pens from A to Z

HowExpert with Lauren Traye

Copyright HowExpert™
www.HowExpert.com

For more tips related to this topic, visit HowExpert.com/fountainpens.

Recommended Resources

- HowExpert.com – Quick 'How To' Guides on All Topics from A to Z by Everyday Experts.
- HowExpert.com/free – Free HowExpert Email Newsletter.
- HowExpert.com/books – HowExpert Books
- HowExpert.com/courses – HowExpert Courses
- HowExpert.com/clothing – HowExpert Clothing
- HowExpert.com/membership – HowExpert Membership Site
- HowExpert.com/affiliates – HowExpert Affiliate Program
- HowExpert.com/jobs – HowExpert Jobs
- HowExpert.com/writers – Write About Your #1 Passion/Knowledge/Expertise & Become a HowExpert Author.
- HowExpert.com/resources – Additional HowExpert Recommended Resources
- YouTube.com/HowExpert – Subscribe to HowExpert YouTube.
- Instagram.com/HowExpert – Follow HowExpert on Instagram.
- Facebook.com/HowExpert – Follow HowExpert on Facebook.

Publisher's Foreword

Dear HowExpert Reader,

HowExpert publishes quick 'how to' guides on all topics from A to Z by everyday experts.

At HowExpert, our mission is to discover, empower, and maximize everyday people's talents to ultimately make a positive impact in the world for all topics from A to Z...one everyday expert at a time!

All of our HowExpert guides are written by everyday people just like you and me, who have a passion, knowledge, and expertise for a specific topic.

We take great pride in selecting everyday experts who have a passion, real-life experience in a topic, and excellent writing skills to teach you about the topic you are also passionate about and eager to learn.

We hope you get a lot of value from our HowExpert guides, and it can make a positive impact on your life in some way. All of our readers, including you, help us continue living our mission of positively impacting the world for all spheres of influences from A to Z.

If you enjoyed one of our HowExpert guides, then please take a moment to send us your feedback from wherever you got this book.

Thank you, and we wish you all the best in all aspects of life.

Sincerely,

BJ Min
Founder & Publisher of HowExpert
HowExpert.com

PS...If you are also interested in becoming a HowExpert author, then please visit our website at HowExpert.com/writers. Thank you & again, all the best!

COPYRIGHT, LEGAL NOTICE AND DISCLAIMER:

COPYRIGHT © BY HOWEXPERT™ (OWNED BY HOT METHODS). ALL RIGHTS RESERVED WORLDWIDE. NO PART OF THIS PUBLICATION MAY BE REPRODUCED IN ANY FORM OR BY ANY MEANS, INCLUDING SCANNING, PHOTOCOPYING, OR OTHERWISE WITHOUT PRIOR WRITTEN PERMISSION OF THE COPYRIGHT HOLDER.

DISCLAIMER AND TERMS OF USE: PLEASE NOTE THAT MUCH OF THIS PUBLICATION IS BASED ON PERSONAL EXPERIENCE AND ANECDOTAL EVIDENCE. ALTHOUGH THE AUTHOR AND PUBLISHER HAVE MADE EVERY REASONABLE ATTEMPT TO ACHIEVE COMPLETE ACCURACY OF THE CONTENT IN THIS GUIDE, THEY ASSUME NO RESPONSIBILITY FOR ERRORS OR OMISSIONS. ALSO, YOU SHOULD USE THIS INFORMATION AS YOU SEE FIT, AND AT YOUR OWN RISK. YOUR PARTICULAR SITUATION MAY NOT BE EXACTLY SUITED TO THE EXAMPLES ILLUSTRATED HERE; IN FACT, IT'S LIKELY THAT THEY WON'T BE THE SAME, AND YOU SHOULD ADJUST YOUR USE OF THE INFORMATION AND RECOMMENDATIONS ACCORDINGLY.

THE AUTHOR AND PUBLISHER DO NOT WARRANT THE PERFORMANCE, EFFECTIVENESS OR APPLICABILITY OF ANY SITES LISTED OR LINKED TO IN THIS BOOK. ALL LINKS ARE FOR INFORMATION PURPOSES ONLY AND ARE NOT WARRANTED FOR CONTENT, ACCURACY OR ANY OTHER IMPLIED OR EXPLICIT PURPOSE.

ANY TRADEMARKS, SERVICE MARKS, PRODUCT NAMES OR NAMED FEATURES ARE ASSUMED TO BE THE PROPERTY OF THEIR RESPECTIVE OWNERS, AND ARE USED ONLY FOR REFERENCE. THERE IS NO IMPLIED ENDORSEMENT IF WE USE ONE OF THESE TERMS.

NO PART OF THIS BOOK MAY BE REPRODUCED, STORED IN A RETRIEVAL SYSTEM, OR TRANSMITTED BY ANY OTHER MEANS: ELECTRONIC, MECHANICAL, PHOTOCOPYING, RECORDING, OR OTHERWISE, WITHOUT THE PRIOR WRITTEN PERMISSION OF THE AUTHOR.

ANY VIOLATION BY STEALING THIS BOOK OR DOWNLOADING OR SHARING IT ILLEGALLY WILL BE PROSECUTED BY LAWYERS TO THE FULLEST EXTENT. THIS PUBLICATION IS PROTECTED UNDER THE US COPYRIGHT ACT OF 1976 AND ALL OTHER APPLICABLE INTERNATIONAL, FEDERAL, STATE AND LOCAL LAWS AND ALL RIGHTS ARE RESERVED, INCLUDING RESALE RIGHTS: YOU ARE NOT ALLOWED TO GIVE OR SELL THIS GUIDE TO ANYONE ELSE.

THIS PUBLICATION IS DESIGNED TO PROVIDE ACCURATE AND AUTHORITATIVE INFORMATION WITH REGARD TO THE SUBJECT MATTER COVERED. IT IS SOLD WITH THE UNDERSTANDING THAT THE AUTHORS AND PUBLISHERS ARE NOT ENGAGED IN RENDERING LEGAL, FINANCIAL, OR OTHER PROFESSIONAL ADVICE. LAWS AND PRACTICES OFTEN VARY FROM STATE TO STATE AND IF LEGAL OR OTHER EXPERT ASSISTANCE IS REQUIRED, THE SERVICES OF A PROFESSIONAL SHOULD BE SOUGHT. THE AUTHORS AND PUBLISHER SPECIFICALLY DISCLAIM ANY LIABILITY THAT IS INCURRED FROM THE USE OR APPLICATION OF THE CONTENTS OF THIS BOOK.

COPYRIGHT BY HOWEXPERT™ (OWNED BY HOT METHODS)
ALL RIGHTS RESERVED WORLDWIDE.

Table of Contents

Recommended Resources 2
Publisher's Foreword.. 3
Introduction 11
Chapter 1: History of Fountain Pens and Regional Variations 12
  The Beginning of Fountain Pens 12
    Lesson 1: What is a fountain pen? 12
    Lesson 2: The Start of Fountain Pens 12
    Lesson 3: The Downfall of Fountain Pens 14
  A Modern Resurgence 15
    Lesson 4: Fountain Pens Make a Return 15
    Lesson 5: The Internet's Role 16
    Lesson 6: Vintage vs. Modern Pens 16
  Regional Variations 17
    Lesson 7: Fountain Pens in the U.S 17
    Lesson 8: Fountain Pens in Asia and Europe 18
  Chapter Review 19
Chapter 2: Anatomy of a Fountain Pen 20
  Body Parts of a Pen 20
    Lesson 9: The Basic Parts of Fountain Pens 20
    Lesson 10: The Cap 21
    Lesson 11: The Body 22
    Lesson 12: Nibs 24
    Lesson 13: Feeds 25
    Lesson 14: Grip 27
  Materials Used and Why 28
    Lesson 15: Fountain Pen Body Materials 28
    Lesson 16: Resin Pens 28
    Lesson 17: Metal Pens 29
  Tips for Picking Your Ideal Pen 30

Lesson 18: Things to Consider 30
Lesson 19: The Price 30
Lesson 20: The Use 32
Lesson 21: The Personal Factor 34
Chapter Review 34
Chapter 3: All About Nibs 36
Nib Sizes 36
Lesson 22: Nib Tip Sizes 36
Lesson 23: Nib Size Numbers 36
Lesson 24: Nib Size Variation 37
Lesson 25: Western vs. Japanese Nibs 38
Lesson 26: Major Nib Brands 39
Lesson 27: Proprietary Nibs 41
Nib Materials 42
Lesson 28: Steel Nibs 42
Lesson 29: Gold Nibs 43
Different Uses for Different Nibs 44
Lesson 30: Flex Nibs 44
Lesson 31: Stub and Italic Nibs 45
Chapter Review 47
Chapter 4: Know Your Filling Systems 48
Cartridge Converter 48
Lesson 32: What are Cartridge Converter Pens?
.... 48
Lesson 33: Standard vs. Proprietary 50
Lesson 34: Cartridge or Converter? 51
Lesson 35: Filling a Cartridge Converter Pen 53
Piston and Vacuum Fillers 56
Lesson 36: What is a Piston Filler? 56
Lesson 37: How to Fill a Piston Pen 57

Lesson 38: Special Considerations for Piston Pens 58
Lesson 39: Vacuum Fillers 59
Lesson 40: How to Fill a Vacuum Filler 59
Lesson 41: Sealing the Ink Reservoir 62
Eyedropper Fillers 63
Lesson 42: What is an Eyedropper Filler? 63
Lesson 43: Pros and Cons of Eyedropper Fillers 64
Lesson 44: Filling Designed Eyedropper Fillers 64
Lesson 45: Eyedropper Conversion 65
Chapter Review 68
Chapter 5: How to Write with a Fountain Pen 69
Writing with a Fountain Pen 69
Lesson 46: Getting Words on the Page 69
Lesson 47: Writing with a Fountain Pen as a Leftie 70
Lesson 48: Tips for Underwriters 70
Lesson 49: Tips for Side Writers 71
Lesson 50: Tips for Over Writers 71
Lesson 51: Tips for all Lefties 72
How Writing with a Fountain Pen Differs from a Regular Pen 72
Lesson 52: The Unique Experience 72
Lesson 53: The Smoothness 73
Lesson 54: At Home Nib Tuning 74
Lesson 55: Getting the Smoothest Experience .. 74
Lesson 56: The Care 75
Tips and Tricks to Get the Perfect Writing Experience 76
Lesson 57: How to Get You Perfect Writing Experience 76

Chapter Review 77
Chapter 6: It's About More Than the Pen; Ink and Paper 79
    Why the Ink and Paper are So Important 79
        Lesson 58: The Fountain Pen Trifecta 79
    Differences in Ink 80
        Lesson 59: Wet vs. Dry Ink 80
        Lesson 60: Shading Inks 81
        Lesson 61: Shimmer Inks 82
        Lesson 62: Best Practices for Shimmer Ink 83
        Lesson 63: Sheening Inks 85
        Lesson 64: Special Considerations for Sheening Inks 86
        Lesson 65: Stubborn Inks 87
    Determining the Right Paper 89
        Lesson 66: Paper Use 89
        Lesson 67: Paper Weight 90
        Lesson 68: Paper Color 90
        Lesson 69: Coated Paper 91
        Lesson 70: Texture 92
        Lesson 71: Ruling 93
        Chapter Review 96
Chapter 7: Get the Most Out of Them: Cleaning and Maintenance 98
    How to Clean a Fountain Pen 98
        Lesson 72: Cleaning a Cartridge Pen 98
        Lesson 73: Cleaning a Pen with a Converter ... 102
        Lesson 74: Cleaning a Piston Filler 103
        Lesson 75: Cleaning a Vacuum Filler 106
        Lesson 76: Cleaning a Converted Eyedropper Pen 107

Lesson 77: Cleaning an Intentional Eyedropper Pen 108
Taking Care of Your Fountain Pen 109
　Lesson 78: Pen Storage on The Go 109
　Lesson 79: Pen Storage at Home 110
　Lesson 80: Pen Storage on Your Desk 110
　Lesson 81: Keep your Pen Capped 111
　Lesson 82: Keep your Pen Boxes 111
Regular Pen Maintenance Tips and Tricks 112
　Lesson 83: Tips for Regular Pen Maintenance. 112
　Lesson 84: When to Clean Your Pen and Refill 113
　Lesson 85: When to Clean and Fill Your Eyedropper Pen 115
　Lesson 86: Pen Flush 115
　Lesson 87: When to Use Stronger Cleaners 116
Chapter Review 116
Chapter 8: The Benefits of Using Fountain Pens 118
The Collecting Aspect 118
　Lesson 88: Fountain Pen Collecting 118
　Lesson 89: Ink Bottle Collecting 119
　Lesson 90: Ink Sample Collecting 119
　Lesson 91: Ink Swatching 120
Artistic and Physical Benefits 122
　Lesson 92: Art and Fountain Pens 122
　Lesson 93: Physical Benefits and Comfort 123
　Lesson 94: Ecological Benefits 123
　Lesson 95: Sanitary Benefits 124
Suggested Uses to Get You Started 125
　Lesson 96: Lyrical Journaling 125
　Lesson 97: Gratitude Journaling 125
　Lesson 98: Memory Keeping 126

Lesson 99: Daily Reflection 126
Lesson 100: Handwriting Practice 127
Lesson 101: Correspondence 128
Lesson 102: Note Taking................................... 129
Chapter Review... 130
Frequently Asked Questions 132
1. What pen should I start with? 132
2. Should I start with cartridges or bottled ink? 133
3. Which way should I hold the pen? 133
4. How important is the paper? 134
5. What does ink with shimmer mean? 135
6. What does ink with sheen mean? 135
7. What does bulletproof ink mean? 135
8. Why are fountain pens so expensive? 136
9. Should I invest in a more expensive pen upfront? 136
10. Why would I want more than one fountain pen? 138
11. My pen won't write. What do I do? 139
Conclusion 141
About the Expert 142
Recommended Resources 143

Introduction

Fountain pens are a unique tool. In today's day and age, no one is using a fountain pen because they have to. They use them because they want to. There are so many alternatives to using fountain pens that they seem like a remnant of the past. But today, there is a bright and bustling community of users who treasure these seemingly forgotten gems.

This book will cover everything you need to know about using fountain pens. You'll learn the differences between the most popular pen types and how to choose the best pen for you. You'll learn not only how to write with a fountain pen but also how to clean and maintain your fountain pen so that it lasts a lifetime.

Fountain pens are not like your average pen. You don't only use them until the ink dries up, or they get lost in your bag. They work to last, and you can use them countless times to ensure you feel that personal touch every time you pick it up.

If you're interested in learning all you can about fountain pens, look no further. You've come to the right place. Let's get started.

Chapter 1: History of Fountain Pens and Regional Variations

The Beginning of Fountain Pens

Lesson 1: What is a fountain pen?

By definition, a fountain pen is a writing utensil that has an internal reservoir that holds water-based ink and uses a metal nib to transfer that ink to the paper through capillary action. Some say that a fountain pen is no more than a controlled leak. One would think that the ink would simply flow right out of the pen, but the flow is easily maintained through the internal mechanisms.

Lesson 2: The Start of Fountain Pens

Historians can date some of the earliest evidence of a device similar to the fountain pen back to Leonardo da Vinci and the Renaissance. Researchers have found drawings in his journals that seem to depict devices that look like pens. These designs showed pens working through gravity or capillary action.

However, the pens that our modern fountain pens most closely resemble were invented and patented in the early 1800s. With the Industrial Revolution's help, they were able to be manufactured in the mid to late 1800s. John Scheffer invented the first design to receive some form of success. He called his invention the 'Penographic' pen and was half feather quill and half metal.

Throughout the late 1800s, cheap steel nibs were easier to manufacture with the help of new machinery, and more designers were patenting pen designs every year. With the aid of new material to tip the steel nibs, a much harder metal called iridium, and the invention of free-flowing ink, fountain pens were steadily becoming the standard writing instrument. They were preferable to dip pens as they were less messy. A user no longer had to have an open ink well out on their desk; all of the ink they'd need was already in the palm of their hand.

Around 1870, mass-produced pens were becoming more popular and easier to come by. Many brands that are still very prominent today made their entrance into the fountain pen world around this time. The largest brand in the U.S that emerged at this time was Waterman. Most of Waterman's pens were 'twist fillers,' or what we call piston fillers today. They did have a problem with these pens leaking, and that was when brands like Conklin made their way to prominence with their 'self -fillers' or what we call a bladder or squeeze filler today. For years, it was a constant battle to have the most popular pen that fixes every user's problem with their instrument. Every solution to a problem seemed to create its own set of issues.

Safety pens became very popular because of their increased protection against leaking. Waterman's safety pen introduced the idea of a screw-on cap with an inner sleeve to keep the nib sealed off when not in use. These two innovations are extremely popular with modern fountain pens. Other safety pens had the ink reservoir wholly sealed off when not in use. With a twist of the pen's body, the nib would retract down into the ink reservoir, and the cap would seal the whole mechanism like a cork, effectively making it safe from leaks.

Lesson 3: The Downfall of Fountain Pens

Fountain pens maintained their dominance as the primary writing utensil used much through the 1950s, soon to be unseated by ballpoints. The original ballpoints were very expensive, and it didn't make sense for many people to switch. But by the 1960s, ballpoints had begun to take over as they were much more manageable than filling a pen with bottled ink.

Fountain pen ink cartridges were becoming popular in Europe. They managed to fight the rise in ballpoint usage, but the convenience of a disposal pen was too great, and fountain pens became more novelty than an everyday use item. In addition to the inconvenience, fountain pens were getting too expensive. Because fountain pens can last beyond disposable ballpoints, pen makers manufactured them with better, more sturdy materials that cost more to produce. As manufacturers switched from hard rubber materials to mediums like acrylic, resin, and celluloid, the pens became more of a luxury item than an everyday use item.

In a matter of a few decades, fountain pens went from the gold standard in writing to collector's items, especially in the United States, where local manufacturers were making pens increasingly more expensive. Many countries were still creating cheap pens across the world that everyone could use and required elementary schools to use them when learning to write. But for the United States, fountain pens became a figment of nostalgia and more of a commodity than anything else.

A Modern Resurgence

Lesson 4: Fountain Pens Make a Return

For decades fountain pens remained relics of a long-gone era. Still, with the advent of the internet, new users are discovering fountain pens and the love of the personalization it brings in a technological age.

Fountain pens are supremely tactile. For many users, it's hard to decide to buy a pen without trying it out first. Some brick and mortar stores that have been around for 70 or more years, like Fountain Pen Hospital in New York, New York or Dromgoole's in Houston, Texas, have stood the test of time and kept their doors open for fine writing enthusiasts to come and discover their newest purchases first- hand. However, not everyone has a store close to them. Thus, they rely on the content put out by online retailers across the country to help them make that decision.

Lesson 5: The Internet's Role

One such retailer, The Goulet Pen Company, took advantage of the internet and started a YouTube channel to educate new and potential users about fountain pens and all of their benefits and provide accurate information about their products' size and performance.

With this new access to information and a generation nostalgic for the past before the internet and the impersonal nature of texting and e-mail, a modern resurgence of fountain pen usage began. People fell in love with handwriting again, and many begun to learn to write in cursive and improve their handwriting now that they had these fine writing instruments in their hands.

Enthusiasts were discovering for the first-time pens that had been around for decades. One of the most popular fountain pens today, the Lamy 2000, was invented in 1966, just as fountain pen usage was dwindling worldwide.

Lesson 6: Vintage vs. Modern Pens

One advantage the modern fountain pens have over their predecessors is the price. Due to advancements in manufacturing, pens can be made far cheaper than they used to, which makes them more collectible. It's rare to find a fountain pen enthusiast today who only has one pen; they all tend to be collectors. Whereas having the best and most prestigious fountain pen used to be the goal, nowadays, it's more about your collection size than what it consists of.

Just because many modern pens are inexpensive, that does not mean that there are no more collector quality offerings. For as many pens collectors quickly purchase without much thought to the price, manufacturers are creating pens that are indeed works of art. Many brands like Pilot and Platinum, both based in Japan, create fountain pens made with special techniques called Maki-E that take months to complete one single pen. These pens can sell for hundreds and sometimes thousands of dollars due to the labor that is involved. Collectors of all caliber can find something in the current fountain pen market.

Collecting is a very modern ideal in the fountain pen community. With the advancements in manufacturing technology that keep costs low and the variety of colors and finishes that many pens come in, writing with fountain pens is no longer necessary; it's for fun!

Regional Variations

Lesson 7: Fountain Pens in the U.S

Throughout the world, fountain pens are very different. The United States manufacturers tend to be on the smaller side and offer more customized and bespoke options. They pride themselves on their handcrafted pens with a personal touch in each pen.

Despite the personalization of the bodies of the pens, many local manufacturers outsource their nibs. European brands such as Bock and Jowo are used universally by pen makers all over the world.

Lesson 8: Fountain Pens in Asia and Europe

Nibs across the world can vary in size and performance despite having the same size marker on them. German or Italian made nibs tend to perform on the broader side than Japanese or other Asian made nibs. Characters used in Japanese writing are more intricate and compact, thus requiring a finer line. Nibs marked as Fine or Medium from a Japanese brand will perform more like an Extra Fine or Fine, respectively, from a German-made nib.

In many cases, the pen is a reflection of the culture. Italian manufacturers create their pens to mimic important works of art from their respective regions and aim to make them perform wonderfully and look beautiful in a pen case or on a shelf. Japanese pen brands tend to favor efficiency and universality. Not to say that there are not truly beautiful Japanese made pens, but most that you will encounter are neutral or muted in color, where their western counterparts will be more vibrant or vary in color. It's all relative to the culture.

While collecting fountain pens is very common in the United States, across the world, children are taught to write with fountain pens from the start. As little to no pressure is required to write effectively, it teaches children not to bear down heavily as they write, and it helps them improve their handwriting. Children in European countries like Germany and Italy are among the cultures that start their elementary-age children with fountain pens specially designed for children.

Chapter Review

1. Fountain pens are pens with an internal reservoir of ink drawn onto the paper from a metal nib using capillary action.

2. Fountain pens were the predominant writing instrument until the ballpoint became cheaper and more accessible due to its disposability.

3. Fountain pens gained traction with modern collectors because of the internet and the ability to spread knowledge and love for the hobby more easily.

4. In different regions of the world, pens are made differently. The United States is home to more small, bespoke pen manufacturers, while Asian and European pen makers have larger manufacturing facilities.

5. Throughout Asia and Europe, children learn to write with fountain pens from the start. While not bearing down hard on the nib, they can write more easily.

Chapter 2: Anatomy of a Fountain Pen

Body Parts of a Pen

Lesson 9: The Basic Parts of Fountain Pens

Fountain pens are both intricate and simple. Regardless of the style you're using, every pen has a few of the same parts that are important to know.

Common Fountain Pen Body Parts

1. Cap
2. Body
3. Grip Section
4. Nib
5. Feed
6. Filling System (this will be covered in a later chapter)

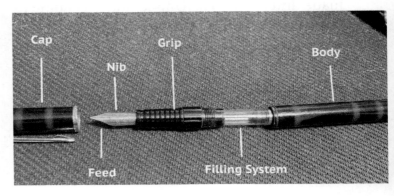

Every pen you encounter has these six body parts. Some of these may seem fairly self-explanatory, but every part is important to understand your fountain pen.

__Lesson 10: The Cap__

The cap and the body of the fountain pen are what make up the majority of your pen. The cap is the part of the pen that closes it off and seals the nib. The body is the part that holds the filling mechanism and sometimes acts as the ink reservoir itself.

Not every cap is created equal. There are snap caps that attach to the body with a push or click, and there are screw caps that attach to the body with a few turns following tracks of threads.

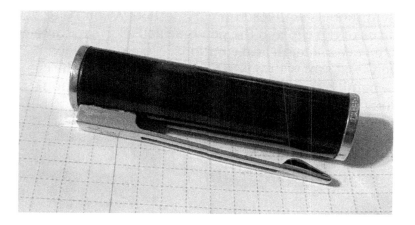

Pen Cap

For some people, the cap type is important. Not everyone has the patience for a screw cap, while others don't like snap caps because they don't seal as well. Think about the purpose you have for your pen. If you value convenience, a snap cap will be more up your alley. If you value a good seal on your pen, you may want a screw cap or a pen with an inner sleeve. Some pens have a sleeve inside the pen's cap to aid in sealing off the air and keeping the pen from drying out.

Brands like TWSBI and Platinum design their pen caps to have this inner sleeve so that the nib can stay wetter for longer. Air is the enemy of fountain pen ink. Since the ink is water-based, it can evaporate very easily. A proper seal in your pen's cap is necessary if you don't plan to use up the whole reservoir of ink within a few days.

For this reason, many pen brands prefer to design their models with a screw cap method. A couple of turns of the cap and the pen is sealed. Pen cap threads can vary in length and rotations. Most commonly, pens will seal within two to three rotations of the cap. Generally, the more a pen costs, the fewer rotations you'll find on a screw cap.

Lesson 11: The Body

Pen bodies are important because they make up most of the pen that you'll be holding. When deciding on a pen, it's important to determine if the pen body's material and weight will be too heavy or light for you.

As the pen body is the largest single piece of the pen, this section will best show off the pen's style; aesthetics do matter. A fountain pen is something you will be using every day. It has to be something you're happy to look at.

The shape of the body can vary from pen to pen. Most are a traditional round shape, devoid of any edges. This is more comfortable for most users as the body is the part of the pen that rests against your hand as you write. While the edges, or facets, carved into the bodies of some pens are not sharp or harsh in any way, some people are sensitive to those kinds of edges.

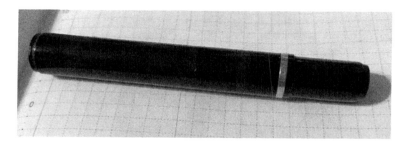

Pen Body

The more work goes into the making of a pen, the more it's going to cost. Smooth, rounded pen bodies tend to be on pens that cost less. It doesn't take as much detail to mold or shape those, while pens with more facets or edges cost more because they have to be carefully carved from the base material.

Some pen bodies are just hollow tubes that cover up the converter or cartridge that holds the ink, but in the case of a piston, vacuum, or eyedropper filled pens, the body is also the ink reservoir. Piston and vacuum filling pens house the mechanism that draws ink up into the pen and seals it off. In eyedropper pens, the body of the pen holds just the ink. A more in-depth description of these filling mechanisms will be available in a later chapter.

Lesson 12: Nibs

The nib is the most important part of the pen and lends itself to the most variation. The nib is what allows the pen to make contact with the paper and write. Fountain pen nibs are always made of metal, most commonly steel, gold, and occasionally titanium or palladium.

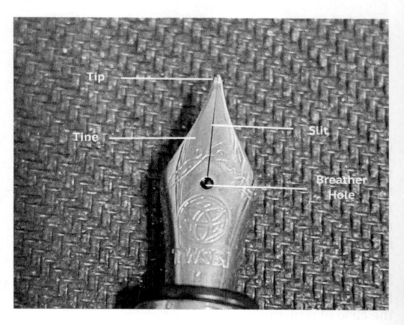

Anatomy of a Nib

The nib is split into a few simple parts: the tip, the tines, the slit, and the breather hole. All of these parts work together to draw ink from the plastic feed down to the paper.

The tip of the pen is made of a different material than the rest of the nib, and normally it is made of iridium, a precious metal alloy that is very strong and will not erode over time, so your nib will always write the way it is meant to. The majority of the nib is made of either steel or gold.

The slit is the line cut up the top of the pen that separates the end into two tines. This slit aids in capillary action and gives the ink a track to travel up as it moves to the tip and down to the paper. For most nibs, the tines are very stiff; they don't move or spread out. Gold nibs are generally softer than steel nibs; thus, their tines have an easier time spreading apart while you're writing and can give you a variation when you're writing.

The breather hole is more for aesthetics in modern pens. It is the hole that rests on the top of the nib about halfway down. They act as an endpoint for the slit so that every nib is cut correctly. In many modern pens, their nibs no longer have breather holes as they don't affect the nib flow anymore.

Lesson 13: Feeds

The feed of the fountain pen is the plastic piece that rests under the nib. The fins and channels draw ink from the reservoir and deliver it to the nib through capillary action. Traditionally feeds were made of black rubber like material called Ebonite. Today, some feeds are still made out of this material, but more often than not, you'll find feeds are made out of plastic. Ebonite feeds are primarily on more expensive pens as the material is more fragile than molded plastic.

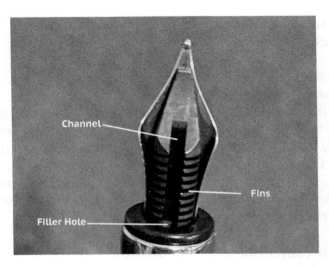

Pen Feed

The feed of the fountain pen is the plastic piece that rests under the nib. The fins and channels draw ink from the reservoir and deliver it to the nib through capillary action. Traditionally feeds were made of black rubber like material called Ebonite. Today, some feeds are still made out of this material, but more often than not, you'll find feeds are made out of plastic. Ebonite feeds are primarily on more expensive pens as the material is more fragile than molded plastic.

Lesson 14: Grip

Lastly, the grip section. This is the place of the pen where you will hold your pen. Grips are important because the more comfortable a grip is, the more likely you will use your pen. Important things to note about a grip are the shape, the size, and the threads.

Depending on the size of your hand, you'll want to pick a grip to match. If you have larger hands, you'll want a wider grip. Alternatively, if you have smaller hands, you might want a thinner grip. Pay attention to how you hold your pens now. Do you use three fingers or four? Do you hold your index finger and thumb close or far apart? Some grip sections are triangular, and if you hold your pen with four fingers, this could be uncomfortable.

Pen Grip

Materials Used and Why

Lesson 15: Fountain Pen Body Materials

The material can make a big difference in how your pen feels in your hand. Some materials are heavier, some are lighter, some may have an adverse effect on your skin, and some are integral to how the pen works.

The most common fountain pen body materials are resin, aluminum, and brass. Generally, the cheaper the pen is, the cheaper the material is to produce. Plain colored resins and aluminum tend to be cheaper because they are easier to cast and inject mold. They can be mass-produced more easily and thus will normally be cheaper than more intricately made pens.

Lesson 16: Resin Pens

Resin is a very variable material. There are a lot of ways it can be turned into your new favorite pen. Injection molding is the easiest and cheapest way to use it. However, resin pens cost more than metal body pens solely because of the craftsmanship required to carve the pens to the exact measurements.

Resin with many swirls and color variations is often individually carved to ensure that the variation is right throughout the pen. Many pens are carved to have faceted bodies or bodies with flat sides all around the pen's circumference, and that takes skill to carve them to just the right width and angle. The labor required is enough to justify the elevated cost of these pens.

Based on this, one would think that clear resin would be cheaper, but that's not always the case. Pens made up of clear resin can be more expensive than you'd expect because of the labor required to make the resin clear. Removing all signs of manufacturing and machining can be a more time-consuming process than some companies are willing to do, so the pen is designed to look cloudy. In that case, the polishing simply isn't needed. Pens with this clear, polished resin are called Demonstrators. Clear models of pens were once made for shop owners to demonstrate how a particular pen worked to their customers. Now they're amongst one of the most popular categories of fountain pens.

Lesson 17: Metal Pens

Metal pens can come in a variety of types and sizes. Before buying a pen made of metal, make sure you don't have any sort of allergy to any metal. Those with nickel allergies may sometimes have reactions to metals in certain pen bodies. If you have some kind of metal allergy, you should be fine if the pen is coated with a lacquer or finish. It's best to look at the manufacturer's specifications or retailer's website or contact the retailer directly to be on the safe side.

Metal is always going to be more expensive than a comparable resin pen. There is more material to work with, and more care needs to be taken with it. All of that factors into the cost. A Kaweco Al-Sport, which is made of aluminum, is nearly triple the price of a Sport pen made of resin.

Metal is also much heavier than resin. This is an important factor to keep in mind when weighing different options. If your hands are smaller, a lighter pen may be best for you. If you are used to bigger pens with more heft, metal could be the way to go. The best part about fountain pens is that it's all up to personal preference.

Tips for Picking Your Ideal Pen

Lesson 18: Things to Consider

When deciding what fountain pen you should be buying, it's important to consider three key factors: the price, the use, and the personal factor. Selecting a fountain pen is a big decision. More than likely, this is a pen that you'll be using for a long time, and you'll be using it every day. It takes time to consider if a pen is right for you.

Lesson 19: The Price

Price, for many people, is the ultimate factor when deciding to buy a fountain pen. When compared to most disposable pens, the price for baseline models can seem astronomical. You can get a disposable fountain pen for about 4 dollars or less. But once you've tried that and want to get into more pens, then what? You have to decide what price is worth it to you.

The beginner price range for fountain pens is between five and fifty U.S Dollars. That's the range where most people starting out will go. At the lower end, you'll find disposable pens like the Pilot Varsity, the Jinhao 993 Shark, and the Platinum Prefounte. The first being a disposable pen, while the other two are cheap but well-made cartridge converter pens. You're only going to find cartridge converter pens at this price point. If you want a pen with a built-in filling system, you'll have to go to the higher end of the beginning price range.

Around thirty dollars is where the quality of the pen starts to increase. The pens don't feel as cheaply made, and you begin to explore metal-bodied pens. The Pilot Metropolitan is a commonly used starter pen because of its great value for its price, ranging between fifteen and twenty dollars. It comes with a cartridge of black ink and a converter so that you can start right away with bottled ink. It writes like it should cost more than it does.

Beyond that, pens with built-in filling systems start to arrive at this price range. The TWSBI Eco is one of the most popular beginning piston filling pens because of its collectability, special edition colors coming out every year, and its price, sitting at just around thirty dollars USD. It may be one of the cheapest piston filling pens on the market, but it's so named for its economical nature. It may not cost you all that much money, but it's worth every penny and then some. If you want a bigger ink capacity and a more comfortable pen to hold, this is a great place to start.

The next level price range encompasses pens that cost between fifty and two hundred dollars USD. You'll find pens made with higher quality materials, gold nibs, and even some handmade pens in this range. Most people don't start in this range before they start using fountain pens. But as you use your inexpensive pens and learn what you like, more pens in this price range may start to interest you.

Beyond two hundred dollars, there is a diminishment in your return on investment. The more you spend on a pen does not always correlate directly to the pen's quality and the experience you'll have with it. A pen is not necessarily better just because it is more expensive. This range often encompasses limited edition pens manufactured in small quantities, pens made of special materials that cost more to produce or take more care to create, or pens that are works of art in pen form. This price range is for the collector. The average fountain pen user may never have a pen in this price range. The price range you stick to is entirely up to you.

Lesson 20: The Use

Decide what you're going to be using your pen for. Are you going to be carrying it with you all day to take notes on the go? Are you going to be using it for school? Are you going to use it for letter writing? How you're going to use your pen is a large factor in deciding the perfect one for you.

For every day carry and use, you may want a smaller pen that can slip into your pocket or clip onto your wallet. You'll want a pen with a finer nib because the finer the line, the less ink is put down, and the faster it will dry. You may also want a pen with extra protection against leaking, like vacuum filling pens with a sealable ink chamber.

If long writing sessions for school are your intended use, comfort is going to matter most. Lightweight pens are easier to write with for long periods because they cause less fatigue. But if you prefer a heavier pen, make sure the pen is the right length. A heavy pen can be comfortable as long as it's not too long. With these, you can go for any nib size and any ink. These notes are for you, so pick something you think your handwriting will look nice and legible with.

Letter writing is very popular among fountain pen users. Not many people get written correspondence in the mail anymore. But with fountain pens, many people decide to bring back the antiquated method of communication. It's a chance to put a smile on someone's face and use a favorite ink. For this, the only factor to consider is comfort and what looks nice to you. If something is pretty, you're more likely to use it. You can use any nib you want. You can try using a stub nib or a flex nib for an extra flourish in your handwriting. Some inks shift color with the light or have pieces of glitter in them. The best thing about picking a pen for its aesthetics is that it's entirely up to you.

Lesson 21: The Personal Factor

This will differ from person to person, as the title suggests. Nothing will be so important as the pen that speaks to you on a personal level. A pen could be perfect on paper, be in the right price range, be the right weight and nib size, but you might not write with it if the pen doesn't appeal to you visually. People are very visual. A pen that looks aesthetically pleasing to them is one they will reach for again and again. If you think a pen is ugly, the odds of you using it goes down drastically.

And that's the best part of fountain pens! There's so much variety to choose from that a pen will appeal to every kind of person. There are pens for people who like bright colors, people who prefer minimalism, and everything in between. Fountain pens are different from any other pen you buy in the store because this is your pen. It's one thing that makes fountain pens fascinating; they are as unique as those who use them. Chances are, very few people out there will be using the same pen, nib, and ink combination as you. It's entirely your own.

Chapter Review

1. The fountain pen comprises six simple components: the cap, the body, the grip, the nib, the feed, and the filling system.

2. The nib is the part of the pen that creates contact with the paper. But it is not the only

aspect in play when getting the ink onto the paper.

3. Most commonly, you'll come across pens made entirely out of resin/plastic or metal.

4. Determine what you're going to use the pen for when deciding to buy it, as comfort and ease of use will factor into the size of your pen.

5. Your pen choice is your pen choice. It's entirely personal and will always come down to what you find most appealing.

Chapter 3: All About Nibs

Nib Sizes

Lesson 22: Nib Tip Sizes

Fountain pen nibs are the most important part of a pen. There are so many options and so many variables that go into the perfect nib for you. Let's start with the basics. What are the fountain pen nib sizes?

The general range of nib sizes will go, from smallest to biggest, Extra Fine, Fine, Medium, and Broad. These will be your most common sizes for regularly ground nibs. These are nibs that are tipped with a ball of iridium and ground to a specific fineness.

Outside of that range, some brands will offer sizes like Ultra Extra Fine or even Double Broad. The most common nibs outside of the normal range are known as Stub nibs. These are special because they're not ground to write on rounded balls of material. Stub nibs are ground square with angles, so your writing will have a unique look as you write. You can denote the different sizes of Stub nib by a number, 1.1, 1.5, 1.8, etc., as this correlates to the width of the nib end in millimeters.

Lesson 23: Nib Size Numbers

In addition to tip size, some nibs are accompanied by a number. #4, #5, and #6 are the most common numbers used to describe fountain pen nibs' length. While the number is not universal, the rule of thumb is, "the bigger the number, the bigger the nib." It's not a perfect system, but it'll get you as close of an idea as you can get about what the pen will write like.

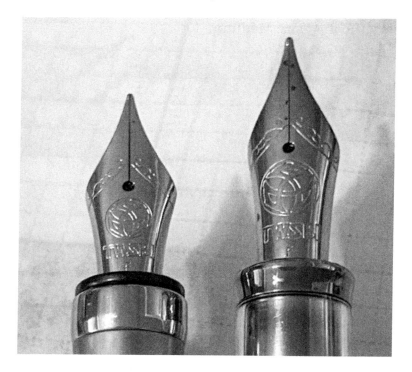

A #5 nib and a # 6 nib

Lesson 24: Nib Size Variation

Because all nib sizes are not created equal, there is some variation in how they write. A Fine nib is not going to write the same across different brands. In some cases, it won't write the same across the same brand.

As the nibs themselves get bigger, so do the tips. A Fine on a #6 size nib will not write like a Fine on a #5 size nib. Because the #6 is larger, it will write much closer to a #5 size Medium.

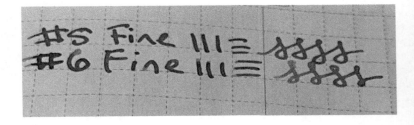

#5 fine vs #6 fine

Lesson 25: Western vs. Japanese Nibs

In fountain pens, nib sizes are not standardized. It makes things more complicated when deciding on the nib size that is best for you. The biggest difference is between Western and Japanese nibs. These nibs are ground with their specific society and writing style in mind.

Western-made nibs, being manufactured likely in Italy or Germany, will write broader and wetter than their Japanese counterparts. A Western Fine may write more like a Japanese Medium. Adversely, a Japanese Fine might write close to a Western Extra Fine. This variation is important when deciding what nib to write with.

But why are Japanese nibs so much smaller than others? Traditionally, nibs manufactured in Japan and other Asian countries are ground smaller because their writing requires a finer line. The characters in the different writing styles found in Asian countries are more intricate than the western style alphabet. A finer nib will give you the precision you need to create the characters legibly without the lines bleeding together. This is also another reason why finer nibs are better for people with small handwriting.

Lesson 26: Major Nib Brands

While nibs are not standardized, a few companies manufacture many nibs for brands worldwide that don't make them themselves. This does make it a bit easier when deciding on the nib that is right for you. If you have experience with the nib brand, you'll know how the pen will write going into the purchase.

The most common companies who manufacture nibs for other pen companies are Bock and Jowo, with the latter being more prolific across the world. Most retailers will have a section in the pen technical specs to tell you if the nib is made by either of these brands or if they're made in house by the pen manufacturer.

If you already have a pen-like a TWSBI with a Medium Jowo #5 nib, you know that if you buy a Kaweco Sport that uses the same nib, you will write very similarly. There is a little differentiation when it comes to the pen's feed, but there will be few differences between how the nibs perform.

Many companies choose to outsource their nib production for a few reasons. Many newer companies that have emerged in the last fifty years don't have that history of nib making some of the original one hundred or more year old companies do. It's not worth developing a system to manufacture their nibs when focusing their efforts on the pen itself. If they don't have to make their nibs because these other companies exist, why should they? It allows them to produce higher quality pens with high-quality nibs with little extra cost.

Some companies will outsource some of their nibs and not others due to a special property they are not equipped for. Italian pen manufacturer Visconti used to outsource their highest-end nib made of Palladium to Bock because the material is complicated to work with. They needed a maker with more expertise to make the nibs perfectly. They've since removed that nib from their line, and now all of their nibs are made in house.

Lastly, a reason that a company may outsource its nib manufacturing is that it's cheaper. Very often, outsourced nibs tend to be on pens in the under two-hundred-dollar USD price range. If a company does not have to front the manufacturing cost, store materials, and develop quality assurance processes, the savings are passed onto the consumer. Put plainly, it's cheaper to have another company manufacture the nibs and more worth it to put those funds into the rest of the pen. This ensures that the consumer gets a higher quality pen for a lower cost.

Lesson 27: Proprietary Nibs

Many brands, especially those who have been around for nearly one hundred years, make their nibs. They have complete control over the process, the shape, the functionality, and the quality. Pen brands worldwide manufacture their nibs for these reasons, giving many of them their unique qualities.

Sailor is a popular pen brand based in Japan. They have been manufacturing fountain pens since 1911. They've become known in the fountain pen community for their gold nib pens with unique tip sizes like Music and Zoom that write differently depending on your writing angle and the slight amount of feedback you get from writing with these nibs. They're able to have better control over these special properties because they make their nibs themselves.

Many Japanese brands, like Pilot and Platinum, are known for making their nibs. Both of these companies are over one hundred years old and have always made their nibs. They have complete control over the quality and the grind of the pens, and it helps their pens be consistent from one to the next. And they use the same nibs over multiple pens.

For example, if you have a Platinum Prefounte, you already know how the Plaisir will write. They use the same nib. In the case of some Pilot pens, like the Kakuno and Metropolitan, you can even swap out their nibs between the models.

Amongst Western Manufacturers, Visconti, Lamy, and Pelikan are a few of the most prolific brands who manufacture their nibs. In the case of Visconti, this can mean that their nib sizes don't match up to any others with comparable nib sizes. Just because a Visconti nib is a Fine does not mean it will write like a Fine Jowo nib on a Kaweco. Visconti nibs are only comparable to each other.

In the case of Lamy and Pelikan, they manufacture their nibs and imbue them with unique properties. Lamy designs most of their pens and nibs with the intention of the nibs being swappable between models. Except for a few higher-end models, you can swap nibs between them with ease to change up your writing experience. You can even buy their nibs separately from their pens, so you don't have to pay the full price for a whole new pen if all you want is a new nib size. Pelikan pens and nibs are little works of art. By manufacturing them in-house, they ensure consistent quality both in performance and looks of their nibs.

Nib Materials

Lesson 28: Steel Nibs

The two most common materials used to manufacture nibs are steel and gold. They each have their pros and cons. They each bring something new to your writing experience and keep you on your toes.

Steel nibs are the most common as they are the cheapest to manufacture and the easiest to work with. Steel nibs are not exactly created equal. While they are on the cheapest four-dollar pens, you may also find steel nibs on more expensive pens where the cost derives more from the body than the nib itself. Some fountain pen users feel that a pen should never have a steel nib over a certain amount of money, but there are exceptions, which differs for everyone.

Steel is the best for beginner fountain pen users. It tends to produce a stiffer nib, one where there is very little bounce or softness to the nib, which is easier for new users to control. The steel in fountain pens is sturdy. And will take quite a bit of abuse while you are still learning how to use the pen. Unless the pen you've purchased is a flex pen, however, you shouldn't press down too hard. A well-tuned fountain pen should write perfectly fine under its weight without any pressure. But if you do, the steel can stand up to it with little damage.

Lesson 29: Gold Nibs

Gold nibs are a great place to go once you've learned what you like out of fountain pens. Gold nibs are softer, thus requiring even less pressure than steel nibs. This softness also allows you to get a little bit of variation when you write if you happen to press down a little bit on your downstrokes. Due to their price, gold nib pens are not recommended for beginners. Nearly all gold nib pens start at roughly one hundred dollars USD, and that is quite a bit to pay for a first fountain pen.

Gold nibs can write on the wetter side when compared to their steel counterparts. Due to the material's softness, they can put down quite a bit more ink than others. The difference isn't as wide as, say, a #5 nib to a #6, but there is a noticeable difference when comparing the nib sizes of gold and steel nibs of the same manufacturer in comparable sizes.

Different Uses for Different Nibs

Lesson 30: Flex Nibs

Beyond the usual Fine, Medium, etc. nibs, some nibs are designed for specific purposes. One such type is a flex nib. As the name would suggest, these nibs are designed so that as you write, the tines flex, creating a larger line than a straight nib.

Flex nibs are often made of steel and can run every end of the price range. Many pens come with flex nib options, like the Conklin Duragraph, known as the Duraflex when the nib is attached, but others are made specifically for flex writing, like the Noodler's Ahab. The only difference is the feed. A pen designed specifically for flex writing will have a feed that is made specifically to deliver as much ink as possible to the paper. When writing with a flex nib, you need more ink to keep that capillary action flowing correctly. If there isn't enough ink, you could end up with skipping or blank spaces inside of your letters.

This type of flex writing is very popular with calligraphers, artists, hand letterers, or anyone who likes an extra flourish in their writing. While you could write a whole letter or diary entry with this type of nib, it's more practical to use it only for your signature at the end of your letter.

Writing with a flex nib is not the same as writing with a regular pen. Because the use of the nib is to bend, you have to put pressure on the pen to have the nib work properly. As you move the pen, press down on the nib to splay the tines carefully. Don't press too hard, or you will spread them out too far, and they won't return together. It takes a lot of practice, but once you've gotten the hang of it, the results are beautiful.

Lesson 31: Stub and Italic Nibs

Stub and Italic nibs are a great way to get line variation without having to add extra pressure to your pen. These nibs are ground so that their ends are flat across, stub nibs having slightly more rounded corners, and italics having sharper corners. Because of this flat edge, you will naturally get wider downstrokes and thinner cross strokes.

A 1.1 Stub nib

Many people like to use these in their everyday writing. These nibs are much more like the average fountain pen nib and thus are easier to control than flex nibs. Stub nib sizes will run slightly larger, but that's the only difference from a regularly rounded nib. Some users prefer the look of their handwriting with this line variation, and they like the added wetness the larger nib gives their writing. This lets them appreciate the color or special properties of their ink better, and thus they enjoy writing more.

You can get an italic grind on smaller nib sizes. Some custom nib smiths will do this for you, although they may not provide this service for nibs smaller than a medium. There comes the point when there is simply not enough material to work with. They can grind a Japanese Extra Fine nib down to an italic grind, but with a line that small, you may not be able to see the variation at all. It's best to have these types of grinds on larger nibs so you can appreciate the line variation.

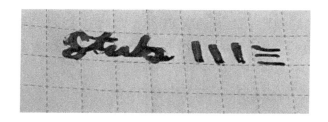

Stub writing

Chapter Review

1. Nibs primarily come in sizes extra-fine, fine, medium, and broad.
2. Nibs manufactured in Asian countries will always run smaller than nibs manufactured in European countries due to the written languages' different requirements.
3. Most commonly, nibs will be made out of steel or gold, with the latter being more expensive.
4. Flex nibs are ideal for practicing calligraphy and other styles of hand lettering.
5. Stub and italic nibs can be used for calligraphy and lettering but are more like traditional nibs, so they're easier to use.

Chapter 4: Know Your Filling Systems

Cartridge Converter

Lesson 32: What are Cartridge Converter Pens?

Cartridge Converter pens are the most common types of pens you will encounter. Nearly every brand of pen has one of these in their line. Cartridges are little tubes of ink that you can stick into the back of your pen to write with. These are very popular because they are the most convenient method of getting ink into your pen. You don't have to mess with a bottle or any other kind of tool. You can also keep extra cartridges on you in case you run out while you're writing on the go.

Ink cartridges are very popular in European and Asian countries. As such, many pens will come with one ink cartridge to get you started if you're not quite ready to delve into the world of bottled ink.

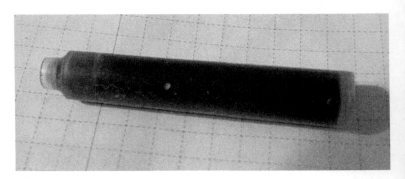

Standard International Short Cartridge

In the United States, bottled ink is more popular. Many people who are pen collectors are also ink collectors. There's a wider variety of ink to choose from in bottled form, and more ink with special properties tend to come only in bottles. As such, we need ways to use our cartridge pens with bottled ink. That's where converters come in.

As the name would suggest, a converter converts your cartridge filling pen into one that can make use of bottled ink. Converters work by a small piston inside the ink chamber. You twist the knob at the back of the converter, and it moves the piston to draw ink up into the pen.

Standard International Converter

Converters add an extra level of maintenance to the pen itself. Instead of simply popping out the empty cartridge and popping a new one in, you need to clean the converter between each new ink you try. Use the converter like you would fill with ink but fill it with clean water and dispel the dirty water into a cup. Repeat as many times as needed until your water runs clear.

Lesson 33: Standard vs. Proprietary

Nibs aren't the only things that manufacturers make themselves. Some make their pens specifically so that they fit only their cartridges and converters. This is from a time when pen manufacturers preferred that their users use only their ink in their pens to ensure that there is no issue with the flow of the pen. Nowadays, as ink production is becoming more standardized and more companies are producing ink outside of pen brands, you can use whatever ink you desire.

But if you want to use a cartridge, you will be stuck using that pen brand's cartridges. A few prolific companies that require proprietary cartridges and converters are Lamy, Pilot, Platinum, and Sailor. You can't use a Pilot cartridge in a Platinum pen. If you want to use cartridges with these brands, you will be limiting yourself to the kinds of ink you can choose. If you want the option to try other inks, just in case, you will need to purchase a separate converter if the pen does not already come with one.

To counteract these proprietary cartridges, many pen brands have started to use a more standardized design where their cartridges sit in the pen's grip. These are known as Standard International. Any pen that can use these cartridges or converters will say so on the retailer's website. Standard International cartridges also come in various sizes, most commonly short and long. All Standard International pens can hold short cartridges, but you'll have to check the technical specifications to see if the pen body is long enough to hold the long cartridge.

Any pen that is fitted with the ability to use a Standard International cartridge or converter will say so in any retailer information. And often, some pens will come with a cartridge and a converter to get you started.

Lesson 34: Cartridge or Converter?

Which should you use? The decision is very personal. Each option has its sets of pros and cons.

Pros of Using Cartridges

1. Convenience. You just pop the cartridge in, and you're good to go.
2. Ease of use. There is zero learning curve when learning how to fill your pen with a cartridge.
3. Overall Price. Packs of cartridges will normally be cheaper than full bottles of ink.
4. Storage. As packs of cartridges are smaller than bottles, they're much easier to store on your desk or the go.

Cons of Using Cartridges

1. Limited options. The variety available in cartridges simply isn't as large as the variety you get when looking at bottled ink.
2. You can't try before you buy. You can't buy a smaller quantity to try out an ink before committing to a whole pack. However, you're

limited to buying the whole pack of cartridges and hoping you like the ink.
3. Smaller ink capacity. A short Standard International cartridge will hold just around .75ml of ink, where a converter, depending on the brand, will get you closer to 1ml. You can get a larger ink capacity with long cartridges, but you will need a pen that can fit the larger cartridge.

Pros of Using Converters

1. Ink Variety. With a converter, you are open to using the vast collection of bottled ink at your fingertips.
2. Larger ink capacity. Most converters range between .8ml capacity and 1ml capacity. Additionally, when you draw ink up through the feed of the nib and into the converter, you have the added capacity of the feed itself, which can add an extra .25 to .5ml of ink to your overall capacity.
3. Using Shimmer and Sheen Inks. More often than not, inks that contain shimmer and sheen don't come in cartridges. By using a converter, you're open to bottled ink and thus more inks with special properties.

Cons of Using Converters

1. More Cleaning. Since you're not throwing out the converter when it's empty, you do have to clean it along with your pen between uses.
2. Slight Learning Curve. Like most pens with an internal filling mechanism, converters take a little time to get used to the right angle and speed to move the piston to get the fullest fill on your pen.
3. Mess. Converters don't always seal perfectly. Sometimes they're not seated correctly, or the piston knob gets turned, and thus you do open yourself up to a higher chance for leaking.

It's up to you to weigh these pros and cons and decide which method is worth it for you. Many people with cartridge converter pens opt to use both because they like the convenience of cartridges with the option to use bottled ink if they want. The decision is up to you.

Lesson 35: Filling a Cartridge Converter Pen

If you've chosen to use cartridges, filling your pen is a simple as unscrewing the grip from the body of the pen and pressing the unused cartridge against the end of the exposed feed until you hear a snap. This snap indicates that the cartridge's seal has been broken, and the feed is beginning to suck up the ink. You may need to wait a few minutes before writing as it will take some time for the ink to travel up the feed channels.

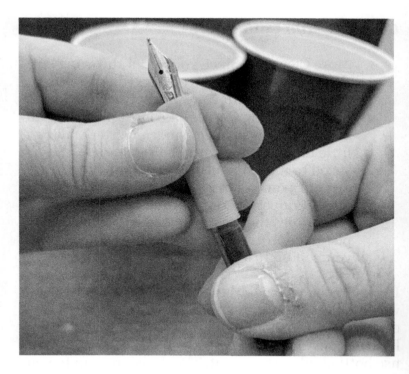

Pressing the Cartridge into Place

Cartridges are traditionally used once and then disposed of, but you can reuse them. When a cartridge is empty, or you've done using the ink, use a blunt tip syringe and flush the cartridge with clean water. Once it's clean, you can use that same syringe to suck up some ink from a bottle or vial and dispense it into your cartridge. Then replace the cartridge, and you're good to go.

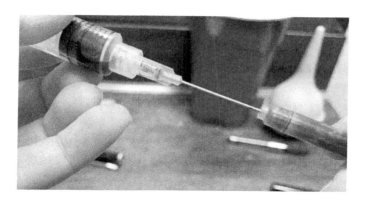

Filling a Cartridge with an Ink Syringe

If you're using a converter, make sure the converter is already attached to the pen's grip and dip the nib into your bottle of ink until the whole nib is submerged, this will ensure ink can get into the filler hole. Slowly twist the knob on the back of the converter to draw ink up into the pen. Go slowly and do this a couple of times to ensure there is no excess air and you are getting the most ink you can into your pen.

Filling a Converter

Piston and Vacuum Fillers

Lesson 36: What is a Piston Filler?

Piston filler pens are fountain pens with a built-in filling system that operates with a piston that moves the plunger inside the ink reservoir up and down to fill the pen with ink. As these systems are inside the pen itself, you don't have to remove the body to get to the filling system. It's all there and ready to go. Additionally, because the filling system is a part of the pen, you can only use bottled ink with these pens; you cannot use cartridges.

A Piston Filler Pen

Piston pens are the most popular filling system for pens with built-in filling systems. Since there is no extra ink reservoir, the whole body of the pen is the reservoir; you get a much larger ink capacity. It's often nearly double that of your average converter. If you mostly use bottled ink and have no use for ink cartridges, piston filling pens may be the way to go if you want to increase your ink capacity.

Lesson 37: How to Fill a Piston Pen

Piston filling pens work by dipping the nib into your bottle of ink up to the filler hole and turning the piston knob at the back of your pen. This will lower the plunger to remove all of the air from the reservoir. Then turn the knob in the opposite direction to raise the plunger. The suction created by this is what draws ink up through the feed and into the ink reservoir. After wiping any excess ink off of your nib and grip section, you are ready to write.

Filling a Piston Filler

Lesson 38: Special Considerations for Piston Pens

Watch out. The initial fill is going to write much wetter than it normally will. The feed is primed with a lot of ink, and the line may be much wider than expected. To avoid some of this, hold the nib over the ink bottle and dispel a few drops out by lowering the piston. Then turn the knob in the opposite direction to raise the plunger and suck in some air. This will ensure that the feed isn't oversaturated and won't write wetter than it's supposed to. Many brands recommend this as a normal step of the filling process because it also decreases the likelihood that your pen will leak excess ink into the cap of your pen or on the page as you write.

With larger ink reservoirs, you do run the risk of some burping. The more air there is in the ink reservoir, the more pressure is created, and occasionally, this pressure causes the ink to erupt from the nib when it's not supposed to. This is more of a problem with vintage pens than with modern pens, but it's something to be cautious of. If you like, as the ink level drops, you can lower your piston plunger, so there is less space for the ink to move around, thus less air.

Lesson 39: Vacuum Fillers

The vacuum filler was invented to counteract some of the issues you get when a pen burps. These pens fill a little differently from piston fillers as they rely on a vacuum pressure to fill the pen with ink.

The telltale characteristics of a vacuum filling pen are a knob that can be pulled back entirely after being unscrewed, a body that flairs slightly as it meets the grip section, and the thin metal rod attached to a vacuum gasket in place of a twist piston.

A Vacuum Filler

Lesson 40: How to Fill a Vacuum Filler

To fill a pen with a vacuum filling mechanism, first, you unscrew the knob at the back of the pen and pull it back to move the gasket to the back of the ink reservoir. Once you've dipped your nib into your pen up to the filler hole in the feed, slowly press the vacuum piston down, and it will slowly increase the pressure inside. Once it's reached the end of the pen body, where the body flairs out slightly, the vacuum seal will break, and the ink is sucked up into the pen via reverse vacuum pressure.

Pulling Back on the Vacuum

Depressing the Vacuum

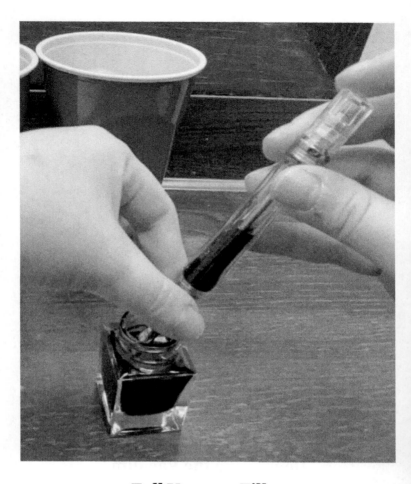

Full Vacuum Filler

Lesson 41: Sealing the Ink Reservoir

Once you have your desired fill, you can twist the knob to seal off the ink reservoir. You will have to loosen this slightly when you want to write for longer periods to let more ink into the feed. But the ability to seal off the ink chamber makes this pen very travel friendly. You reduce the chance of ink burping into the cap during everyday travel, and as long as you keep the gasket sealed, you can even fly on an airplane with this pen inked up without any issues.

Because of the lack of large internal parts, these pens have much larger ink capacities than their piston filling counterparts. With some practice and technique, you can get a full fill on these pens, resulting in an ink capacity of 2.3ml or more, depending on the pen.

Eyedropper Fillers

Lesson 42: What is an Eyedropper Filler?

Eyedropper pens are the descendants of the original fountain pens. Before piston filling mechanisms and ink cartridges were invented, the only way to get ink into your pen's body was to transfer ink from your bottle to your pen using an eyedropper. As such, any pen where you fill the body with ink directly is dubbed an eyedropper pen.

Lesson 43: Pros and Cons of Eyedropper Fillers

These pens are great because of their ink capacity. It's as large as the pen itself. There's no need to clean extra parts like converters or maintain moving parts like pistons. All you need to do is get ink from point A to point B.

However, some eyedropper pens can cause additional issues when it comes to burping. Much like piston pens, when there is a lot of air inside the barrel, it will create pressure that can cause excess ink to leak from the nib into the cap, on your paper, or your hands. Additionally, if the connection point between the grip and the body is not watertight and secure, you can risk ink leaking through the threads and creating a mess.

Lesson 44: Filling Designed Eyedropper Fillers

Due to these complications, beginners who want to explore the eyedropper pen sector may want to start with pens designed to be eyedroppers and have no other filling mechanism. The Stipula Passaporto is a nice pocket pen to start with. Suppose you're willing to spend a little more money. In that case, the entire Opus 88 line of pens are specifically designed to be eyedropper pens and have a sealing mechanism built-in, much like the gasket on vacuum fillers, to reduce burping. Once you fill the barrel of your pen with ink and put the grip section back on, you can raise or lower the rod inside the pen using the knob at the back and seal off the ink chamber, so all you have to write with is what is in your feed. All you have to do is loosen the knob and let more ink up when you need more.

Lesson 45: Eyedropper Conversion

One thing that is very popular in the fountain pen community is to convert a cartridge converter pen into an eyedropper filler. For some, they like the increased ink capacity you get when compared to cartridges. For others, it makes more sense economically.

A popular eyedropper convertible pen is the Platinum Preppy. This pen is five dollars USD and comes with one ink cartridge. Platinum does have a converter that fits the pen to reuse it after you've finished the provided cartridge. However, the converter costs eight dollars USD and thus nearly doubles the cost of the pen. Thus, many people opt to convert the pen into an eyedropper pen to save on costs and drastically increase their ink capacity.

To convert a cartridge converter pen to an eyedropper pen, all you have to do is place an O-ring, pictured below, around the top of the grip section of your pen and place a little bit of silicone grease on the threads to make them more watertight. Once you've done that, use an eyedropper or blunt tip syringe to suck up some ink from your bottle of choice, and deposit it into your pen body up to where the threads start. This will leave room for the top half of your pen and will decrease the chance of ink seeping into the threads and out the pen.

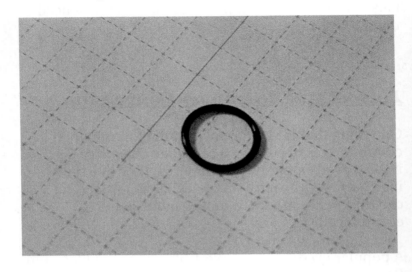

O Ring

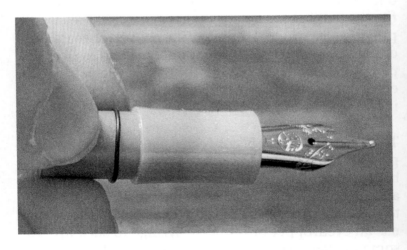

O Ring on the Pen Grip

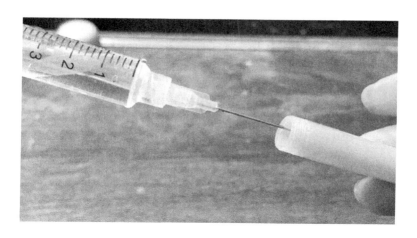

Filling the Body of the Pen

If a pen is eyedropper convertible, some retailers will mark them as such. It's important to know the criteria for a pen that can be eyedropper converted so you don't run the risk of damaging your pen.

Can It Be Converted?

1. Solid-Body. The body of your pen must be one solid piece. Make sure there are no holes, gaps, or ink windows in the pen body. This will cause a leak unless you use some form of epoxy to seal them up.
2. No Metal Parts. While some pens have metal pieces designed to come in contact with ink, most cartridge converter pens that you will encounter and want to convert will not have this treated metal. Untreated metal will rust and corrode if it comes in contact with liquid for too long. A good rule of thumb when deciding if a pen is eyedropper convertible is if

there's metal in the body or on the threads, it's not convertible.
3. Is it worth it? Eyedropper converting is a fun way to switch up the usability of your pen. It can breathe new life into a pen you've had for a very long time. But the risks don't always outweigh the gains. If you're not sure something can be converted, you may damage the pen. If you don't already have the necessary materials, such as silicone grease or an ink syringe, the overall cost may be more than a converter. You have to decide for yourself if this is a worth it venture.

Chapter Review

1. Cartridge converter pens are the most common filling mechanism style and are found on most beginner level pens.
2. Using cartridges is more convenient than converters, but it limits your overall ink choices.
3. Piston and Vacuum fillers are popular due to their larger ink capacity.
4. Pens made as Cartridge Converter pens can be converted to eyedroppers to increase their ink capacity, but this also increases their maintenance.
5. Pens designed to be eyedropper filled have the largest ink capacity but require the most due diligence when cleaning.

Chapter 5: How to Write with a Fountain Pen

Writing with a Fountain Pen

Lesson 46: Getting Words on the Page

It may seem like second nature, but writing with a fountain pen will be unlike anything you've ever experienced before, especially if you're used to ballpoint pens. When it comes to ballpoints, the ink is the gel inside the cartridge, and you need to bear down with a great deal of force for the friction to melt the ink enough for it to transfer to the paper.

With fountain pens, you don't need to exert any pressure at all. You are not forcing ink on the page; you are guiding the nib and letting the pen do the rest for you. Hold your pen lightly in your hand however is comfortable. If you have a three-finger grip, make sure you're not pinching the barrel too hard, as this will cause hand fatigue.

Start with a slight 45-degree angle and touch the nib lightly to the page. The ink will be drawn down through the feed, and your writing will take form. Depending on your style of writing, you may need to slow down a bit. Not every pen is tuned to write quickly, as the feed is not set up to supply the amount of ink necessary for fast writing. Enjoy the experience. Writing with a fountain pen can be a meditative process as well as a tool for communication.

The Ideal Writing Angle

Lesson 47: Writing with a Fountain Pen as a Leftie

It's a common misconception that lefthanded writers can't use fountain pens. Yes, there are some extra things to consider. But lefthanded writers can enjoy the experience as much as righthanded writers. Fountain pens are designed to work in a pulling motion, where lefthanded writers apply a push form of pressure, but there are ways to make sure your writing looks as clean and crisp as possible.

Lesson 48: Tips for Underwriters

If you're an underwriter, meaning you write with your hand underneath your pen, you're in the best spot to avoid issues. Your writing experience won't be all that different from anyone else's. You may need to apply even less pressure than you think to avoid any paper issues as you push the pen to the side.

Lesson 49: Tips for Side Writers

If you're a side writer, meaning your hand is in line with the pen as you write, you are at the greatest risk for ink smearing on your hand. Fountain pen ink dries very slowly, and it's more likely that your hand will run over freshly written words at some point. The best advice for this writing style is to write slowly, write with a drier ink that will dry more quickly, and to use more absorbent paper. The faster the ink dries, the better off you will be.

Lesson 50: Tips for Over Writers

If you're an over writer, meaning your hand is crooked on top of the line as you write, you fit somewhere between the other two styles. You essentially move the pen in the direction it's supposed to be moved, but your arm is in line with you're writing, and thus you can get ink on yourself before it's dried. Ensuring your arm is lifted off the page will help, and the drier ink and more absorbent paper tricks.

Lesson 51: Tips for all Lefties

In addition to these above tips, some brands like Lamy offer left-handed grind nibs. These are nibs that are ground with a sharp edge to help with the push motion of left-handed writers. These nibs will be broader, so if you prefer finer nibs, you may not be able to find a leftie grind on your preferred nib size.

Fast-drying ink will be your friend. If you are using a wetter ink that does not dry quickly, you will want to use a finer nib. Generally, the less ink you put on the page, the faster it'll dry. Conversely, if you want to use a broader nib, you need to use a drier ink.

How Writing with a Fountain Pen Differs from a Regular Pen

Lesson 52: The Unique Experience

Writing with a fountain pen is a unique experience. Each time you write with the same pen can be unique from the last time you used it. Changes in the room temperature or humidity, the atmospheric pressure, or how much sun you've gotten today could change how your pen writes from day to day.

Extra pressure can cause burping, but it can also cause a pen to write much wetter than it normally does. Or on a very dry day, it can write much drier than you anticipated. It keeps you on your toes. For some people, this can be a good thing or a bad thing.

It's almost like the pen is alive. It has preferences, wants and needs, and certain circumstances that allow it to write optimally.

Each pen is different. No two users with the same pen, ink, and paper combination will have the same writing experience due to outside factors. This makes the process of writing with a fountain pen very personal. While every pen is unique, even those made with the same specifications, there is something very special about your personal pen and ink choice.

The likelihood that someone else has chosen your exact pen, nib, ink, and paper combination is very small. This gives your choice a very personal feeling. The experience is entirely your own.

Lesson 53: The Smoothness

As mentioned above, fountain pens don't need any extra pressure to write. You don't need to grip the pen or bear down on the nib. You're encouraged and instructed not to. Because of this, the experience of writing with a fountain pen versus a ballpoint or rollerball is very smooth.

Some nibs are ground specifically to have a hint of feedback as they write, sometimes giving a similar feeling to a graphite pencil. But many nibs are ground with smoothness in mind. It's a very strange and welcoming experience to feel your nib glide across the page with ease after using nothing but ballpoints for years.

Generally, cheaper pens will not write as smoothly as those that cost more money. It takes a lot of time and technique to grind a nib to write smoothly. More often than not, these nibs need to be ground by hand or in smaller batches to ensure that they are formed correctly. When companies manufacture cheaper pens, they don't put in the hours needed for such close attention to detail.

Lesson 54: At Home Nib Tuning

You can make a cheaper nib smoother by tuning it with something called Mylar paper. With ink in the pen, draw a series of figure 8 symbols repeatedly as desired. This files off sharp edges on the nib and smooths out the writing experience. Do this at your own risk. By performing this kind of nib tuning, it is possible to over tune to the point where the nib cannot make proper contact with the paper. Additionally, self-tuning may void your pen's warranty. If there are any other issues with your pen, the manufacturer might not fix or replace it if you've done any tuning to the nib.

Lesson 55: Getting the Smoothest Experience

If you value a smooth writing experience, it may be worth it to invest in smoother paper. The best writing experience will contain your ideal pen, ink, and paper combination in many cases. Brands like Rhodia,

Clairefontaine, and Leuchttrum have paper that is very smooth and fountain pen friendly. A brand of paper, Tomoe River Paper, is the grail of fountain pen friendly paper. It's coated so that it doesn't absorb ink as much as others, which allows you to see your ink's properties. It's very smooth. With the right pen, it can almost feel like writing on glass. It is thin. Paper is measured by its 'grams per meter' or GSM. Tomoe River Paper mostly comes in 52 and 68 gsm, which is very thin for most paper. Due to the thinness, it's very easy to see the ink from page to page. You'll have to decide if this is a problem for you.

Lesson 56: The Care

Fountain pens do require a little more care than regular ballpoints. When you start writing with fountain pens, you'll notice that you use up far more ink than you realize. It becomes the one pen that you use. Fountain pen ink gets used up much faster than your average disposable pen because it's water-based rather than a melted gel. The average note-taker can make one standard international converter last anywhere from one week to one month, depending on how often you use the pen. But if you write bigger or use a broader nib, you'll be putting down more ink with every stroke and thus use up more ink.

Because you're using up your ink more quickly, it could be worth it to keep ink with you on the go in a small sample vial or keep a bottle of your desired ink at your office or on your desk, ready to fill. This fear of running out of ink at the most inopportune time is why many users justify having multiple pens inked up at once. Make it a habit of checking your ink level before you leave if you plan to write a lot out on the go, and your pen doesn't have an ink window for you to check your level as you write.

Because fountain pen ink is water-based, it evaporates and can dry inside the feed, causing issues with your ink flow. Fountain pens require a greater amount of care and attention while writing. You may still have ink in your pen, but the feed could be too dry, dust or other particles could be clogging the feed, or the nib could be misaligned. Good pen maintenance can be tedious, but it's worth it for the enhanced experience.

Tips and Tricks to Get the Perfect Writing Experience

Lesson 57: How to Get You Perfect Writing Experience

1. Pick a comfortable pen. As mentioned in previous chapters, make sure your pen is as comfortable in your hand as possible. When a pen is the right length and weight, it'll be that much more enjoyable to write with.
2. Pick a nib that works best with your writing. Larger writing? Go for a bigger nib. Smaller handwriting? The finer, the better.
3. Pick an ink to match your pen's flow. If you have a smaller nib or a drier feed, you'll get a great experience with a wetter ink. Conversely,

if you're pen is broader, you can use any ink you want. Even shimmer and sheening inks! With those, the more ink you put on the page, the better.
4. Paper does make a difference. For the smoothest writing experience, go for Rhodia or Clairfontaine paper. You will have the smoothest experience with the least amount of ghosting.
5. Go slowly. Write slowly. The more slowly you write, the more you lessen the likelihood that your pen will skip or experience a hard start. Enjoy the experience.
6. Go lightly. Don't grip the pen too tightly, and don't press down on the page too hard. Writing lightly will ensure the best flow possible and the most comfortable writing experience. When you don't grip too tightly, you reduce your hand fatigue.
7. Don't be afraid to try new things. If a certain pen and ink combination isn't for you, it's okay. Try something else. There will always be a new ink to try with your pen or another option to try a new nib. For many people, what's more important than any particular pen is the trifecta of the perfect pen, ink, and paper combination.

Chapter Review

1. Hold the pen comfortably in your hand with the nib facing up. Don't press too hard on the paper. Let the pen do the work for you.

2. Your pen might not write the same every single day due to weather and atmospheric changes. Be prepared to get some ink on you.
3. Write slowly; this will ensure you're getting proper contact with the page.
4. If you're a leftie, take special consideration to how you hold your pen and what ink you use.
5. Writing with a fountain pen should be smoother than your average pen. If it feels too scratchy, adjust your grip and don't press down as hard.

Chapter 6: It's About More Than the Pen; Ink and Paper

Why the Ink and Paper are So Important

Lesson 58: The Fountain Pen Trifecta

As mentioned in the last chapter, what can make or break your fountain pen experience is the perfect combination of the right pen, ink, and paper for your desired experience. This can take some trial and error, but it is more than worth it to make sure that your writing experience is as perfect as possible.

The same ink could look vastly different depending on the paper you write it on. A more absorbent paper can cause the ink to spread and look wider than it's supposed to. Or the ink may feather, causing little feather-like lines to appear out of the ink. It's not ideal for most people. You can counteract an absorbent paper with a finer line. The less ink you put on the paper, the less ink there is to spread and absorb.

Shimmer inks may not look their best on absorbent paper. For the ink to shine, it needs to lay on the paper and dry rather than absorb into it. Not every paper is going to play nicely with every ink.

Depending on your use, it could take some time to find the ink and paper that work best for your pen and your setting. If you're looking to use your pen for work, you may want it to dry quickly, but you still want to appreciate your ink's different properties. It's all about trying new things and finding the combination that fits just right.

Differences in Ink

Lesson 59: Wet vs. Dry Ink

Many times, ink is referred to as a very wet ink or very dry ink. This characteristic is about how it flows through the pen. Different ink manufacturers will put certain compounds in their pen that let them flow quicker or slower. Inks that are wetter flow more quickly and tend to pool up on the page more easily, making them dry more slowly. Inks that are dry flow more slowly. You have to write with them more slowly to get a lot of ink on the page, and sometimes dry inks won't flow properly in broader nibs. These inks are great for some people because they dry more quickly, and they can get on with their writing without risking smearing it all over their hands.

You may want wetter ink if you plan to write with a flex or stub pen. You want to make sure you're putting as much ink on the page as possible with those. While flex nibs do work nicely with dryer inks, wet inks shine most brightly with larger nibs with the right paper. Some wet flowing inks are mixed with a lubricant that allows them to flow through the pen more easily. This compound is often added to ensure that a pen flows without drying and does not stain the barrel of the ink reservoir or converter. If an ink can't stick to the walls of the reservoir, it can't stain the clear plastic. Many people who write with clear demonstrator pens like to use lubricated inks for this reason. It ensures their pen remains crystal clear.

There's a time and a place for all types of inks. It all depends on your use and your need. Like the pen itself, the ink you choose is very personal. There's more to the choice than simply the color.

Lesson 60: Shading Inks

There's something very unique about fountain pen inks. Color variation is a fascinating characteristic that you don't find with other pens. In other pens, when you see the ink become lighter or darker with every stroke, you think something is wrong with the pen. But with fountain pens, it's sought after.

Shading refers to a variation in the darkness of the ink. As you write, some parts of the line will put down more ink, causing a shift from a lighter shade of the ink to a darker one. Inks with low saturation of dye in them are more likely to shade because when there is less dye and less ink, you're more likely to have a greater difference from when you put more ink down.

This color variation is best seen with medium and up size nibs. You can especially get quite a bit of shading if you're writing with print lettering. More ink is left on a letter where you pick up your pen than when you start the line. The more you pick up the pen, the more you'll have these spots of deep color.

If you have an ink that doesn't shade well, and you would like it to, you can desaturate the ink a bit by mixing in distilled water. It's recommended that you do this in small quantities. Put some of the ink into a dish or sample vial to test it out before you commit to diluting the entire bottle. Adding more distilled water will decrease the saturation of dye in the liquid and give you a higher chance of shading.

Most inks will have a chance to be shaders, although you find that the more ink you try, some will shade better than others. Test out new inks in different pens and find what works best for you.

Lesson 61: Shimmer Inks

Some inks come with special properties that go beyond shading. Shimmer inks and sheen inks have become very popular in recent years. These go beyond the color variation you get with a shading ink, one that's still within the same color family, and gives the ink a whole new range of color to work with.

Shimmer inks are called so because they shimmer from the physical glitter particulates in them. Actual shimmering pieces are suspending in ink and give your writing a very magical feeling. Many people like shimmer inks because of the aesthetic. It's very pretty on the page and makes lettering pop. But it doesn't come without its challenges.

Shimmer Ink Swatch

Lesson 62: Best Practices for Shimmer Ink

Because the shimmer is suspended in the ink, it can fall out of the suspension very easily. Most often, if you pick up a vial of ink that contains shimmer, all of the glitter will have settled to the bottom of the container. Before filling a pen with this ink, give the bottle a shake to mix the particulate in with the ink and do so quickly. The glitter will start to settle to the bottom nearly immediately. Don't shake the bottle too vigorously. Shaking too hard causes air bubbles that can get into your pen and disrupt the flow. Just turn it over, while closed, a few times in your hand to disperse the shimmer, and you're good to

As the glitter settles to the bottom of your bottle, it'll do that in a pen that sits still as well. Before using your pen that has shimmer ink, give the pen a few rolls in your hand. Check the converter if you can and see if the ink has been mixed, and start writing. You will have to do this every time you use the pen. The shimmer settles quickly.

While beautiful, shimmer ink is very temperamental to deal with and requires very specific care when cleaning it out. Glitter gets everywhere, and it never really goes away. If you want to start using shimmer inks, you may want to pick one pen and dedicate it to shimmer inks. These ink types require more thorough cleaning than others because the shimmer particulate is very difficult to get out of the pen and the feed. If the pen you're using allows you to remove the feed from the grip section, consider taking it out and soaking it in water for a few hours to loosen the glitter, then take an unused toothbrush and scrub off any particles that are left behind. You may get most of the glitter off, but it'll be hard to get it all.

Because of the glitter pieces, shimmer ink tends to clog up feed channels a lot. As the ink in the feed evaporates, it leaves the glitter behind that blocks up the ink flow. If you're going to use a shimmer ink, use it up as quickly as you can, or don't fill the pen all the way. The longer ink sits in the feed, the harder it dries, and the more difficult it is to clean out.

That being said, shimmer inks are very popular. The glitter in the ink does make writing pop off the page, and they are a lot of fun to play with. If you're not sure you like shimmer inks, see if you can find a sample of the one you're looking for, so there's no need to commit to a whole bottle and purchase a cheap pen that's only a few dollars. That way, if it's hard to clean or it clogs too much, you won't have ruined something you would want to use otherwise.

Lesson 63: Sheening Inks

With some of the benefits of the different color variations you get from shimmer ink without quite as much of the clean-up, you have sheening ink. Sheening inks will consist of a base color that sheens or shifts in color to a different color depending on the light. The most common sheening inks will consist of a red or yellow sheen as these dyes mix the best with other colors.

This sheening effect is caused by special dyes incorporated into the ink. When enough ink is put down on the page, the color variation comes to life. You'll often find blue, teal, and green inks that sheen red and purple or pink inks that will sheen yellow. Many fountain pen users like sheening inks because of the unique color property. It's gorgeous to look at and can make your letters and journaling come alive on the page with the right paper.

Sheen Ink Swatch

Lesson 64: Special Considerations for Sheening Inks

Much like shimmer inks, to get the most color out of your ink, you need to use a less absorbent paper and put as much ink on the page as you can. There are some inks, like Organics Studio Nitrogen or KWZ Sheen Machine, that seem to give off sheen no matter what you write on, but those super sheeners are more difficult to manage than the average sheening ink. If a paper is marketed as a fountain pen friendly paper, it will pull out enough sheen to satisfy you.

In the case of the super sheener inks mentioned above, the extra dye and additives put into the ink make them much wetter and dry much slower than your average ink. This can be an issue, especially when the objective is to get more ink on the page. When the ink doesn't dry, you run the risk of smudging your writing and getting it all over your hands and every surface near you. If you want an ink that will sheen but dries a bit quicker, try Lamy inks that are marketed to sheen and Diamine inks. They won't sheen quite as well as a super sheener, but they will dry much faster so you can get on with your writing.

Lesson 65: Stubborn Inks

For those who don't want special properties in their ink, they don't want color variation. They want the most color for their dollar and some inks are supersaturated with color. But for that much saturation, there are risks. A popular brand of ultra-saturated inks is the Noodler's Baystate series. There are three inks, blue, pink, and purple. These inks have heavy saturation of their dyes and have a slightly more basic pH level than other inks.

This extra saturation of dye and change in pH can make these inks gorgeous on the paper and quite a bit resistant to water and other spills you might get on the page. But these are notoriously hard to clean out of your pen; most highly saturated inks are. They tend to stain the ink reservoir or not clean out of your feed properly. These will need extra care when cleaning if you plan to change the color of your ink afterward.

The staining may not be an issue for you if you're not using a clear demonstrator or your pen has a converter that you can't see without disassembling the pen. But if you want to clean out your pen from these inks, you'll need to beyond the normal soap and water method to clean them out.

Instead of a mixture of a little bit of soap and water, you'll need to flush your pen out with a combination of 10% bleach to 90% water. Do not mix any other cleaners or soaps in with this mixture. Just the bleach and water should be enough. Use this to flush out your pen and feed, or you can let it soak for an hour or so, then rinse thoroughly. You should be able to get most of the ink off as long as it hasn't been sitting in the pen for too long.

For every ink with a unique property, there may be a drawback to consider. You'll have to decide what's important to you, the aesthetic or convenience. And decide what inks are worth the trouble. For some, using bleach to clean out their pens is where they draw the line. So, they have one pen designated to this special ink; that way, they don't have to worry about getting it perfectly clean every time. There will always be ways for you to get exactly what you want out of your pen.

Determining the Right Paper

Lesson 66: Paper Use

Not all paper is created equal, and the choice of paper can make or break your fountain pen experience. The paper is almost as important as the pen itself. You'll never have a smooth writing experience if you're writing on roughly textured paper. Depending on what you'll use it for, your choice of paper may be different.

Are you writing a letter? Do you need the shimmer or sheen of the ink to pop off the page? Are you writing for work and need the ink to dry as quickly as possible? Or are you simply journaling and want your ink to look it's best? These are all things to consider when deciding what paper to buy for your fountain pens.

The most important things to consider when buying paper are:

1. The paper weight is listed as 'GSM' or grams per square meter. In some stores in the United States, you may find paper marked with grams per pound as well.
2. The color. Do you want true white paper or cream paper? The color of the paper will change how your ink looks.
3. Coated or uncoated. Much like the color and the weight, this will change how your ink appears.

4. Textured or Untextured. The more texture you have, the less likely you are to get the smoothest writing experience possible.
5. Ruling. Do you like dots, grids, lines, or nothing?

Lesson 67: Paper Weight

The GSM of the paper you're writing with has a big impact on how the pen writes. The thicker paper has the advantage of less ghosting, seeing shadows of the ink through the page, and less bleed through, ink seeping through completely to the other side. For most fountain pen friendly brands, the paper is going to range between 80 and 120 gsm. This will give you the best chance to show off the ink qualities with little interference.

There are, however, exceptions. Thicker paper tends to be more absorbent. If you use paper higher than that range, you won't have as much ghosting as thinner paper, but it will still interfere with the ink properties. You could have some feathering or spreading as the paper takes in the ink, thus making your writing look different.

Lesson 68: Paper Color

Paper color is entirely preferential. Some people prefer the way their ink looks on certain colored paper over the other. It's something you'll have to try for yourself to find out what works best for you. Most commonly, notebooks and notepads with fountain pen friendly paper will come with white or cream pages.

Some writers want the truest representation of their ink on the page, so they go for the true white pages. For other writers, a true white paper is too bright for them and they prefer a softer undertone to their writing; thus, they write on cream or off-white pages. This cream does shift the ink color slightly, as there is more of a yellow undertone to the page rather than true white. But at the end of the day, it's all about your preference. Choose what looks best to your eye. Think about the notebooks you've used in the past and about what color the pages were. You'll likely find an unconscious pattern of color preferences.

Lesson 69: Coated Paper

Some fountain pen-friendly papers have a special coating on them to make them more ink resistant. This creates a small loophole in the rule that thicker paper is better for fountain pens. Tomoe River Paper is specially designed to bring out the best qualities in the ink that's put on it. The pages are thin, so there is quite a bit of ghosting. But you can pile the ink on without any bleed through and see the best properties of your ink.

You can buy pads of Tomoe River Paper or buy notebooks by other brands that use this paper. Endless Recorder has hardback casebound notebooks that utilize the 68 gsm variety of Tomoe River Paper. If you're in the market for a planner and have a little more money to spend, Hobonichi is a Japanese stationery brand that sells many dated and undated planners that use 52 gsm Tomoe River Paper.

With the special coating, the paper can be very thin without bleeding through, feathering, or spreading. You can pile on your ink to get the most sheen, shade, or shimmer out of it as you possibly can. As the coating makes the paper very ink resistant, you will experience longer dry times than with other papers. If convenience and quick-drying are important to you, you may want to consider thicker paper without this ink resistant coating.

Lesson 70: Texture

Textured paper gives off a unique quality to your writing. Under the ink, you can see the bumps and fibers that make up the page as you write, but this does disrupt the smooth writing experience you get from other papers. The fibers that may get stuck in the pen can impede the flow of ink, thus making them less desirable for fountain pen use. Cheaper papers tend to have more texture than those made for fountain pens.

Paper is smoothed through increased processes of rolling to compress the fibers as tightly as possible. This extra processing is what makes it so expensive. But if the paper does not have to be rolled and pressed quite as many times, then the over-all product will feel rougher but cost less.

If the smoothest writing experience is what you're going for, you'll want to invest in higher quality paper. That's not to say there aren't higher end papers designed specifically to be textured. These can be great for letters or for counteracting extremely smooth pens. Not everyone likes the feeling of writing on glass. If a nib is almost too smooth, you can counteract that with a rougher paper.

Lesson 71: Ruling

This, much like color, is all down to your personal preference and use. No ruling is objectively better than another. It's all about what you like to look at and how you plan to use the paper. The most common ruling types are lined, grid, and dotted. Each has its benefits and uses.

If you plan to do a lot of writing, notetaking, or journaling, lined will be your best option. If the object of the game is to get words on the page, the lines will keep you organized and your writing all in a line. Line widths can vary. Most retailers will list the millimeter width of their ruling, so you can decide what's best for you. If your handwriting is larger, opt for something with a larger ruling or vice versa.

Lined Ruling

Grid lined paper is great if you plan to use the paper for school or work and like to make graphs, charts, or split up your pages a certain way. No matter how you want to organize your paper, there's a line to guide you. Many popular fountain pen friendly paper companies make paper that comes in grid lined notebooks and pads. There are also options within grid lined if you want a bolder grid or a light grid that you barely notice. If there's something you can think of that you want out of your paper, someone's made it.

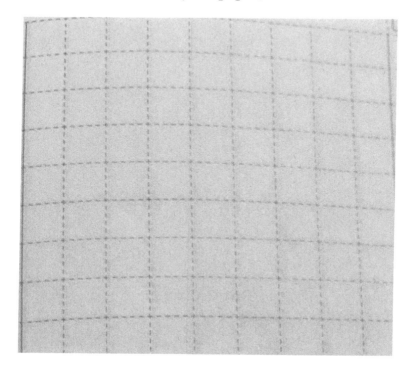

Grid Ruling

Dotted pages are great for bullet journaling, art journaling, or anything in between. The dots are lined up in a grid formation that can guide you when drawing boxes or calendars. But you're not restricted to stay within the lines. Having them lined up as dots gives you more freedom to draw or write as you see fit. Dot grids are a nice middle ground between traditional grid spacing and lined paper.

Dot Ruling

Chapter Review

1. Your ink and paper choices are instrumental in getting to your perfect writing experience.
2. Shimmer and sheening ink will get you the most extreme color variation of any ink.
3. Wetter inks dry slowly while dryer inks dry more quickly.
4. Paper quality will change how your ink looks on the page. Aim for paper weighted between 80 and 120 GSM to be on the safe side. Only use lower GSM paper if it's coated.

5. Paper can have a texture that disrupts ink flow. See if you can try a small quantity of paper before committing to a whole notebook or ream of paper.

Chapter 7: Get the Most Out of Them: Cleaning and Maintenance

How to Clean a Fountain Pen

Lesson 72: Cleaning a Cartridge Pen

Whether you like it or not, cleaning out a fountain pen is part of the use. You have to keep the pen clean so that the ink moves smoothly through the feed and onto the paper. As far a cleaning goes, cleaning a cartridge converter pen has the most options, so you have some wiggle room to decide how you want to go about cleaning your pen.

With cartridge converter pens, you have the option to clean the whole system out as one or separately. Both have their advantages and disadvantages. It simply depends on what you prioritize.

If you choose to flush out your nib section first, remove the pen's body, pull out the cartridge or converter, and run the nib section under the running faucet. This will force clean water through the feed. Keep it there until the water coming out of the nib runs clear, or when you touch the nib to a towel or piece of tissue, there is no ink residue left behind.

Alternatively, you can buy a bulb syringe from a pen retailer or your local pharmacy. They're very inexpensive and hold a lot of water. You fill the bulb syringe with water, hold the end against the back of the grip section where the end of the feed is exposed, and press down on the bulb to force water through. This is a very quick way to clean out your feed. Very often, you can get the whole feed clean with just one fill from the syringe.

Cleaning the Grip Section

Once you've flushed your nib successfully, place the nib, shiny side down, on a towel or other absorbent material to draw the excess water out of the feed. This can mix with the ink you put in and make it write more faintly than you expected.

If you are using cartridges and not refilling them, throw away the used cartridge and pop in a new one. Your pen is ready to go to write once the ink makes its way through the feed. However, you can clean out and refill the same cartridge a few times if you'd like. This can be a great way to try out a bottled ink if you don't have a converter at the ready.

To clean and fill a cartridge, you will need a blunt tip syringe that you can purchase at most pen retailers. The end is flat and large, the needle is not sharp at all, so you don't need to worry about hurting yourself with this. It's a very thin straw attached to a plunger.

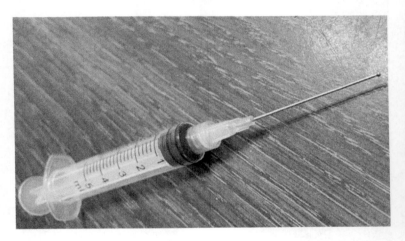

Blunt Tip Syringe

Fill the syringe with water and place the end inside of the cartridge as you depress the plunger to flush clean water into the cartridge. Repeat as many times as needed to clear out any ink residue left in your cartridge. Once it's clean, you can suck up any excess water with the syringe and fill the cartridge with ink from a bottle or vial. Then you pop the cartridge back in, and you're ready to write. The only downside to this is that the plastic ink cartridges are made of can stretch over time, so the seal may not be as tight as it was when you first filled it. Therefore, you should only reuse the cartridges a few times. They're not meant to be reused indefinitely like converters.

Flushing a Cartridge

Lesson 73: Cleaning a Pen with a Converter

If you're using a converter, you can clean out both the pen and the converter simultaneously as long as they're connected. First, remove the pen body and dispel any ink left in the pen into a cup or down the sink. Using the converter, suck clean water from the running faucet or a cup up into the pen and dispel that into your dirty water cup. Do this as many times as needed until the water coming out of your pen runs clear. Then you are ready to fill your pen again.

Water in the Converter

Removing Water from the Converter

If you've cleaned the nib and feed separately from the converter, you can clean the converter with a blunt tip syringe like a cartridge or use the piston mechanism in the converter to flush clean water in as many times as you need.

Lesson 74: Cleaning a Piston Filler

Cleaning a piston filling pen is slightly different from a cartridge converter because you can often not remove the nib and grip section from the rest of the pen. The only way to clean them is to use the piston to fill the pen with clean water and dispel the dirty water into a cup as many times as needed until the water runs clear.

Filling a Piston with Water

Emptying the Piston

Most piston pens cannot disassemble as easily as others, so you have to be more careful with the inks you choose to fill them with. Like TWSBI, some brands make their pens to have the user disassemble them when needed for cleaning and maintenance. Those that are meant to be disassembled will come with instructions and the necessary tools. Sometimes the plunger on the piston no longer moves as smoothly as it did before. This is normal, as the water can wash away some of the grease that keeps the piston moving smoothly against the walls of the barrel.

Follow the directions that come with the pen if it's disassemblable. If not, it's not recommended to try if you are a beginner. Most piston pens are not meant to be fully disassembled. A few flushes with water should do the trick.

Lesson 75: Cleaning a Vacuum Filler

Cleaning vacuum filler pens are very similar to piston filler pens. Use the vacuum mechanism to flush clean water up into your pen. You can give it a gentle shake to release any lingering ink on the walls of the pen barrel and dispel the dirty water out. Repeat this process until the water runs clear.

Cleaning a Vacuum Filler

If your vacuum filler or piston has a nib that can be removed from the pen body, you can clean the nib out with a bulb syringe and the pen body out with a blunt tip syringe like you would with a cartridge converter pen.

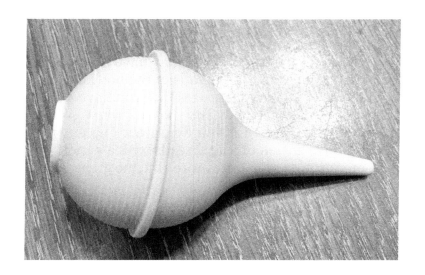

A Bulb Syringe

Lesson 76: Cleaning a Converted Eyedropper Pen

Eyedropper pens are both easier and harder to clean than those above. Because the body and grip come apart, you can flush the nib like you would with a cartridge converter pen. You don't have to worry about cleaning out a cartridge or converter, and there is no piston to keep lowering and raising to get water in and out of the pen to clean it. All you have to do is run the pen's body under the running faucet or flush it with clean water in a blunt tip syringe. But if your pen has a lot of threads, or you're dealing with a converted eyedropper pen, you do have to be very meticulous about getting all of the ink out of every nook and cranny possible.

If you're cleaning a pen that you converted to be an eyedropper, you'll need to reapply your silicone grease to the threads every time. Some may have gotten washed off if it got on the threads of the body of the pen. This silicone grease ensures that you have a watertight seal on the body and that ink won't get out and get on your fingers as you write. Wick up any ink that may have gotten on the threads with a paper towel or cotton swab, apply more silicone grease, then refill your cleaned pen body.

Lesson 77: Cleaning an Intentional Eyedropper Pen

You won't have to worry about applying more grease if you've purchased a pen that was designed to be an eyedropper. In that case, you need to be meticulous about getting all the ink out of the body and off any internal mechanisms before you refill with new ink. Take an ink syringe and flush the body with clean water until no more ink comes back out of the end. Pull back any plungers or plugs meant to seal off the ink chamber.

Once the ink reservoir is clean, you can clean the nib and grip section like a cartridge converter pen. With pens that are designed to be eyedroppers, as they have a valve system to seal off the ink chamber, you don't need to worry about burping as much as a pen you've converted to an eyedropper fill pen. With intentional eyedropper pens, you can use the ink down to the last drop as long as you remember to activate the seal when you're done writing. This also means you have to clean out and refill the pen far less frequently.

Taking Care of Your Fountain Pen

Lesson 78: Pen Storage on The Go

Taking care of your fountain pen is about more than keeping it clean. For a fountain pen to write at its best, proper storage and care are important. Pens can get nicked or dinged when they roll around in your bag, and this excess of movement can force ink out of the feed and into the cap. Keeping your pen safe and secure while you're on the go is crucial. The type of storage you need will depend on your usage, how many pens you have, and what types of pens you have.

If your pens are made of metal or wood, it's best to keep them away from keys or anything else sharp that could scratch the pen's finish. If there's a pocket in your bag where you can keep your pens separate from its contents, all the better.

Make use of your pen's clip. Whether it's in a pocket, a sleeve, or a case, clip your pen into place. It's designed to keep your pen safe and secure while it's stored. This will keep the pen from rolling around too much and building pressure in the ink reservoir.

Lesson 79: Pen Storage at Home

There are pen cases to meet every need you could have. From single pen sleeves to briefcases that hold over 90 pens, you'll be able to find something that fits your needs. A pen case that holds between one to three pens will be a great size to fit in your bag or briefcase for traveling. They'll keep your pens secure and safe while you move and will keep them from rolling away from you when you're ready to write.

If you start to have a pen collection that is getting larger than you know what to do with, brands like Monteverde or Giorlogio make large pen cases to house your uninked pens. While it's not unheard of for people to carry pen cases that hold 30 or more pens with them on the go, they're meant more for home storage. Your pens will be all in one place and be kept safe from dust, dirt, or damage.

Lesson 80: Pen Storage on Your Desk

For desk use, you can get pen-holders of various shapes and sizes. Most are metal or wooden cups with small holes in them for your pen to stand up in. And some are meant to hold one pen as it lies on its side. You can get those in any shape you want, even crabs or other animals!

These are not only aesthetically pleasing, but they make sure your pen doesn't go rolling off your desk. Should that happen and your pen is capped, it is no worse for wear. You may have to clean up any ink spills. Desk holders are especially nice if your pens do not come with clips or roll stops. You don't want to risk them rolling off your desk with nothing to stop them.

Lesson 81: Keep your Pen Capped

By Murphy's Law, anything that can go wrong will go wrong. If you leave your pen uncapped on the table next to you, it's more likely to fall. Due to the way fountain pens are weighted, it's most likely to fall on the ground nib first and render the nib unusable.

You could get lucky and have this happen with a pen that has a nib you can easily replace. But most of the time, dropping a pen like this is the end of that pen's journey. Nib smiths can repair nibs that have been damaged like this. However, most of the time, having the pen repaired is more than the cost of a replacement. If your pen is precious and irreplaceable, always cap it before you set it down. Should the pen fall while it's capped, worse comes to worst, some ink will be forced out through the feed, but the pen will be no worse for wear. You'll have to wipe the excess ink off the grip section and rinse out the cap.

Lesson 82: Keep your Pen Boxes

Except for some cheaper pens that come in plastic sleeves, most pens you encounter will arrive to you in a box or tin of some kind to keep it safe while in transit or to be displayed in a physical shop. It's to your benefit to keep these boxes so you have a place to store your pen when it's cleaned out and not in use. When your collection grows large enough that you don't want to have all of your pens inked up at the same time, you need to have a place to store those pens.

You can find several storage options online to hold your pens. Still, until you find you require a large storage container to house your collection, keeping them in the boxes they arrived to you in is a great way to keep your pens safe from dust, dirt, or damage until you're ready to use them again.

Regular Pen Maintenance Tips and Tricks

Lesson 83: Tips for Regular Pen Maintenance

Bottom line, if you want your pen to write properly, you have to clean it. If you're using the same ink in your pen, you don't have to be as thorough as you would be if you were changing inks, but it's still important to get everything scrubbed out and ready for further use.

The best tip for keeping your pen clean is don't ink it if you're not going to use it. A lot of fountain pen users will ink up a lot of pens because they want to have as many options at the ready as possible. But that can get out of hand very quickly if you don't keep track of the pens with ink in them. Then one day, you'll pull one out of your case to write, and the ink inside will be all dried up in the feed and crusted to the side of the converter. Fountain pen ink is water-based. So, it will eventually evaporate and leave the dye and other components behind if it's left to sit too long in a pen.

Only ink up what you can use within a month. On average, a pen that's in regular use should be cleaned every four weeks or so. A good flush with some clean water will get any extra particles or dust that may have gotten into the feed. If any ink has dried in the channels of the feed, this will also clear those out.

You need to clean your pen out if you want to change ink colors. If you want the truest representation of your ink, you need to clean out the previous color as much as possible. This is not only good from an aesthetic point, but it's the safest way to avoid mixing inks. Many inks are formulated to be as close to neutral pH as possible. But there is always the possibility that some inks will be more acidic or basic. You don't want to find out which ones aren't compatible by mixing them in your pen or converter and having them melt it. It's best to get as much ink out as possible before adding in the new color.

Lesson 84: When to Clean Your Pen and Refill

Most people will use a pen until it's dry and not bother with cleaning out a pen before it's out of ink completely. Others will change their ink weekly or daily regardless of how much they have left. Changing out the ink at the right time is subjective to each pen user. But there are some guidelines when it comes to changing ink solely for maintenance purposes.

If ink has sat in the pen for too long and has evaporated slightly, the ink will become thicker and impede the flow of your pen. It's best to use up your ink before this happens or use a pen designed to keep pens from drying out like this, like TWSBI or certain Platinum pens. If you notice that your ink is coming out darker or in larger pools than normal, it's a sign that it's time for you to give your pen a flush and refill it with new ink.

When pens sit around unused, ink left in the feed will dry and adhere to the feed's plastic, disrupting the flow. If you pick up your pen to write and no ink comes out at all, you can flood the feed by lowering the piston in the pen or converter, but most likely, it's safest to clean out the pen and start fresh.

On average, a pen filled with ink should not go unused for more than four weeks at a time. That gives the ink plenty of time to dry or evaporate and clog up the pen. If you're not changing ink regularly or going through your whole ink capacity quickly, it's best to limit the number of pens you have inked at once to avoid letting any sit unused for lengthy periods.

Lesson 85: When to Clean and Fill Your Eyedropper Pen

If you're using an eyedropper pen, it's recommended that you don't use the ink all the way down to the last drop. By then, there will be too much air in the pen, and it could cause the ink to be pushed out the feed of the pen when you're not expecting it. If you keep your pen filled to just about halfway, you'll reduce the risk of your pen burping. It will require a bit more maintenance, but it's worth it not to have to wash the ink off your hands after every single use.

Lesson 86: Pen Flush

When cleaning a pen that may have been inked for too long, you have options besides flushing with clean water to get the ink out. In many cases, plain water won't do. A common additive to the water is dish soap. If your nib doesn't seem to want to get rid of the last of its ink, fill a cup with room temperature water and stir in just a little bit of dish soap. Drop the whole nib unit into the water and let it soak. You can also fill a converter with soapy water if there is stubborn ink that won't cleanout. Let the nib sit for a little while, for ten minutes to an hour. Time will differ depending on how long the ink had been left in the pen. Check on it by touching the nib to a paper towel and see if any ink is coming off. If the ink is still pulling out of the feed, let it continue to soak. You can leave a nib unit in soapy water up to overnight if it's been extremely stubborn.

If there's an ink that won't be cleaned out, you can buy a pen flush cleaner from some pen retailers that are designed to pull stubborn inks out of pens and feeds. This is usually an ammonia-based cleaner, so do not mix it with bleach. That will create a toxic gas. Mix the pen flush with some warm water and a little dish soap, and give your pen another soak. There are few inks that this will not remove.

Lesson 87: When to Use Stronger Cleaners

Diluted bleach solutions are useful for some very stubborn inks like the Noodler's Baystate series mentioned in a previous chapter. Mixing 10% bleach with 90% water, you can use this mixture to clean out those inks. Bleach mixtures should only be used for inks that refuse to be cleaned out with soap and water. Never mix a bleach dilution with pen flush. The two types of cleaners will create toxic gas, and that can be very dangerous. For 99% of the ink you're going to encounter, this won't be necessary. Warm water and soap do the trick most of the time.

Chapter Review

1. Clean your pens thoroughly when changing ink colors to keep inks from mixing.

2. Use a bulb syringe to flush out your nib rather than twisting your converter knob over and over again.
3. Clean your pens often, even when not changing inks. Cleaning once every four weeks will ensure your pen is writing smoothly every time.
4. Use pen flush mixed with warm soapy water to clean out any stubborn inks. Just leave your nib to soak for a few hours, and the ink should clean out easily.
5. Keep your pen boxes! If you don't have a place to safely store your uninked pens, keep your pen boxes so you have somewhere to put them when they are not in use.

Chapter 8: The Benefits of Using Fountain Pens

The Collecting Aspect

Lesson 88: Fountain Pen Collecting

Using fountain pens is about more than having a nice tool. For many users, collecting fountain pens is as much about the hobby as using them. Finding the last few editions to complete their collection is what keeps them excited about fountain pens. There are many types of collectors. Some are the 'Gotta Catch Them All' type and collect every color of a certain pen model or brand. Others like to have complete sets of pen models and have one of every pen model that a certain brand may come out with. Then some like to have lots of fountain pens. Many of them pique their interest and don't feel the need to stick to one type, color, or model. They like to have a variety of pens that make them happy.

Fountain pen collecting is not cheap, especially if you're someone who likes to collect limited edition or vintage fountain pens. Those can get up in price very quickly. But others are more cost-effective. TWSBI comes out with new colors of their Eco pen every year, so collectors have something to add to their collection. Each pen is about thirty dollars USD. Compared to other companies who release new colors every year, like Sailor, their model the Pro Gear Slim comes out with new editions every year, but to collect them all, you need to spend between 180 and 200 dollars per pen.

You won't know what kind of collector you are until you start to learn what fountain pens you like best. If you're a very practical person, your collection may end up being an expression of that, with your collection only housing a few pens that you use all the time. Collecting fountain pens doesn't mean you have to have a massive collection of pens that needs to be contained in its room of your house. Your collection is all about you.

Lesson 89: Ink Bottle Collecting

Some people collect every pen that catches their eye, and many people find that same excitement with collecting ink. There are so many different ways to collect fountain ink. You can collect by color, property, brand, or just the inks that look interesting to you. For many people, they find this to be a more practical way of collecting within the hobby because they will eventually use up the ink. It makes sense to have extra ink for when you inevitably run out.

Lesson 90: Ink Sample Collecting

When starting out collecting inks, it's a good idea to test out inks before committing to full bottles. Many retailers will sell their ink in sample sizes of a few milliliters so that you can test out the ink. Collecting ink samples can be just as addicting as collecting full bottles. It's fun to test out all kinds of ink to see what's out there. Sampling ink is cheaper than bottles, and it lets you test out more inks. They take up less space and are easier to pick up in larger quantities. Retailers will even sell ink samples in surprise packs so each order can have that extra air of excitement as you never know what kind of ink you'll get.

Lesson 91: Ink Swatching

Part of collecting ink is knowing how the ink performs. Creating a swatch book or guide can be part of the collection. You not only have these beautiful bottles on display, but you have this collection to see how each of the ink colors look so you can choose which one you like best more easily. Creating the swatches and determining what's most important to showcase will be up to you, but here are a few suggestions to get you started.

Suggestions to Start an Ink Swatching Book

1. Use a notebook with paper that you're going to be using most of the time, so you have a way of knowing how the ink performs on your chosen paper.
2. Use a cotton swab or paintbrush to sample ink swatches on the page as evenly as you can.
3. Layer some sections of the swatch to get an idea of how the ink shades or to show off properties like sheen or shimmer
4. List the name of the ink, brand, and if you have a bottle or sample. Sometimes inks will behave slightly differently from a sample as opposed to the full bottle.

5. Swipe a wetted cotton swab across a section of the swatch to test the ink's water resistance.

Ink swatching and testing is a great way to get better acquainted with your ink and will help you learn what you want out of an ink. Additionally, collecting ink is more cost-effective than pen collecting. Your average ink sample will run from about $1.50 USD to $2, and a bottle, depending on the size, will cost between six- and fifty-dollars USD.

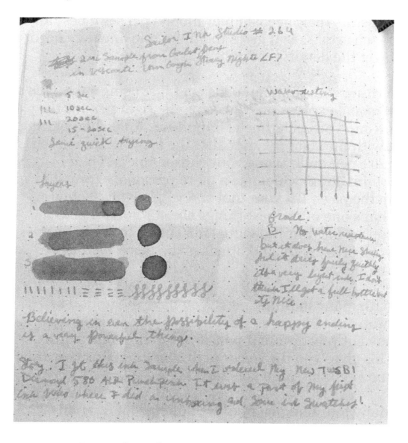

Sample of an Ink Testing Page

Artistic and Physical Benefits

Lesson 92: Art and Fountain Pens

Fountain pens are great for artists as well as writers. Many sketchers and doodlers like to use fountain pens because they can find an instrument that works best for them, and it gives them the freedom to use any ink that suits their needs. Finer nibbed pens are best for sketch artists, as it gives them the fine line they need to get the details just right.

Writing with fountain pens isn't just using a tool. It's a personal expression. Choosing the right pen for the job and you each time is important. And when pairing it with the right ink, it makes the artwork that much more personal. Using your chosen ink and pen combination, you make a more personal expression on the page. It's your tool and your material choice. That's one of the best things about fountain pens. You've got your individual ink choice to create that work of art with, and it makes the piece even more unique.

With many fountain pens, you have the option to swap out the nibs or nib units, so you don't have to buy entirely different pens for different line widths. This gives you the freedom to change size as you draw and keep the same pen and ink.

Many fountain pen inks are bulletproof or archival quality. This means that the ink is permanent. It should not fade over time and is water-resistant. This means that your artwork will last far longer than other fineliner pens or artist pens on the market. The drawing is unique and will withstand anything you throw at it.

Lesson 93: Physical Benefits and Comfort

One of the biggest benefits of using a fountain pen is the lack of hand cramps. As mentioned in previous chapters, writing with fountain pens requires way less pressure than your average ballpoint or rollerball pen. You hardly need to apply any pressure at all to get the pen to write. You're not forcing ink onto the page; you're guiding the pen and letting the ink release naturally from the feed. Without this added pressure, the fatigue on your wrist is greatly reduced, and you can write or draw for longer sessions.

Because of the variety in shape, size, and material of fountain pens, you can find a pen that is the perfect fit for your hand. You no longer have to pinch pens that are too thin for your grip or fumble with others that are too wide. There are enough pens of every size to find the one that fits and weighs just right for you.

Lesson 94: Ecological Benefits

The physical benefits don't stop with you. Using fountain pens over disposable ballpoints is good for the environment as well. You're no longer using disposable plastic that will get thrown away when finished, or worse, lost somewhere and left to the elements. Fountain pens are meant to be reused over and over again. This means less waste from the pack of pens you were using. If you use the same pen and bottle of ink, that one combination could outlast hundreds of ballpoints depending on how long you write with it.

Lesson 95: Sanitary Benefits

For some, this may be a benefit; you'll never have to borrow other people's pens. If you become a true fountain pen fanatic, using other types of pens will soon become undesirable. Nothing compares to the experience of writing with your fountain pen. This may lead to you carrying at least one fountain pen with you everywhere you go and thus never needing to borrow someone else's pen. This allows for less cross-contamination with other's germs and keeping you safe from illness.

Granted, fountain pens are not built for every job. It is nearly impossible to write on the thermal paper used for receipts with a fountain pen. The page is too glossy. You may have to use the pen provided or bring your own as part of your everyday carry arsenal in those cases.

Suggested Uses to Get You Started

Lesson 96: Lyrical Journaling

One of the most common ways to get started using fountain pens is journaling. It gives you the freedom to play with your fountain pen and discover what you like about writing with it. It's your journal, so do what you want with it. Not every entry has to be like a 'dear diary' description of your day. There are so many ways to put pen to paper in this mindful practice.

Transcribe poetry or song lyrics that are meaningful to you. This is a nice way to express your emotions without having to recount those exact feelings into words. You could pick a theme for the month and transcribe songs that fit that theme.

Possible Transcription Themes

1. Songs that are stuck in your head
2. Specific genres such as all rock songs, pop songs, show tunes, etc.
3. Songs from a specific decade or time in your life
4. Songs that evoke a specific emotion

Lesson 97: Gratitude Journaling

Another option for journaling is gratitude journaling. Gratitude entries don't have to be long or overly detailed. Each entry could be one thing you were grateful for that day. This is a nice practice because it gets your writing every day with very little pressure. It doesn't require a lot of time, and it can lift your mood if you've had a particularly bad day.

Lesson 98: Memory Keeping

Memory keeping is a popular method of journaling, especially for parents. Much like gratitude journaling, you don't have to write a lot to make the entry meaningful. One memory of something significant that happened that day is all you need. Some parents like to do this for their children. They'll have one notebook per child and write in the notebook about one thing the child did that day that was special or unique. They do this, intending to give their children the notebooks when they get older.

Lesson 99: Daily Reflection

In the same vein of gratitude and memory keeping, one or some lines a day are popular ways to journal in a low effort method. Some paper companies even make journals and notebooks with this intention in mind. The notebooks have space for anywhere from three to five years, all on one page, so you can see what your life was like on that exact day in previous years. Boiling down your day to just a few lines can help process the events of the day. And it can be fun to see what ink and pen you were using as you go forward in the journal.

Lesson 100: Handwriting Practice

Something many people end up doing when starting to use fountain pens is to improve their handwriting. There is something about the nib's smoothness that encourages users to try using cursive writing again after many years, or in some cases, for the first time. The pen writes very fluidly, and some people think their handwriting looks better with these pens.

If you're learning cursive for the first time or want to experiment with a new writing style, you can use your fountain pen for just that. It's best to start by printing out samples of a font in the style you want to try and emulate with your handwriting. You can either trace this font onto the page or do your best to mimic it with your pen. With handwriting practice, practice makes better. The more you practice, the better your muscle memory will be to create this font style without needing the example sheet beside you.

First, start by copying the letters of the alphabet in both upper and lower case. Try to aim for a few lines of each, if not a whole page, if you're up for it. Copy these letters over and over again until you no longer need the guide sheet to look at. By then, your hand will know how to make the letters the way you want.

The goal of handwriting is to create muscle memory and get your letters to be as even as possible. They should be spaced evenly throughout the line, and they should be about the same size from one letter to the next. Repetition and practice will make it better. The more you do it, the better you'll get.

Handwriting practice is a great use to determine if your pen is comfortable in your hand. You'll be writing with it for whole pages and longer periods overall. If you pair this with a thinner paper, it'll make it that much easier to trace the initial sheet so you can get the lettering down.

Practice all kinds of fonts and find one that feels natural to you. Practice print and cursive writing and determine what you like the most about your handwriting. The type of font you use can also change depending on the size of your nib. If the letters have small flourishes on the ends, you don't want to be using a very broad nib. Experiment with the right nib, ink, and paper combination that will make your handwriting shine.

Lesson 101: Correspondence

Most people don't know when the last time was they got a handwritten letter. With texting and email, it's easier to correspond digitally. It's quick, easy, and always legible. But a handwritten note gives new weight to your words. It says to the other person that you took the time to sit down and write your thoughts on a piece of paper for them.

Writing letters to someone else can be a great way to use your fountain pens. It lets you show off your new hobby with a friend or loved one and sent them a little surprise. It doesn't have to be a long letter. You could write a little 'thinking about you note.' If your recipient is so inclined, you might get a note or letter back and have a new pen pal.

Don't treat the letter like a text or an email. Use the paper to tell a story, tell the other person how you feel, what made you think of them, and ask them questions. Pick your favorite paper to write the letter that you know will showcase your chosen ink the best and help the ink pop off the page.

Now is the time to be a little show offish. If you've been looking for a chance to use shimmer or sheening ink, do it now with this letter. Have fun with the letter and share your enthusiasm for this new hobby with your friend.

Lesson 102: Note Taking

If you're a student or you write down notes for work, or you like to take notes while reading a book, consider taking those notes with a fountain pen. Like handwriting practice, this is a great way to test how comfortable your pen is and how well you can write with it for longer periods. Additionally, it's a great way to get to know your chosen ink and paper combination.

You'll be writing notes that need to be read over again afterward, so take your time when writing to make sure your ink has time to dry and doesn't smear before you turn the page. Write as neatly as you can and learn how your handwriting looks with this nib. Taking notes while reading a book is a great way to get this sort of practice when you don't have school or work that requires you to take notes.

This is the time to test your practical ink choices. See just how fast that fast-drying ink works. Test out as many different black-, blue-, and blue-black inks until you find the appropriate one you like the most. Does your bulletproof ink spread when you use it on the paper you have now? You may have to use a more ink resistant paper when you write with that ink. Writing with these pens and ink as much as possible is the only way to know just how the ink behaves. Do all the testing you can.

Chapter Review

1. Collecting both ink and pens is a fun way to keep the fountain pen hobby new and exciting. This also helps you experience different aspects of finding what you like the most.
2. Writing with fountain pens helps relieve handwriting fatigue as you are not gripping the pen tightly or pressing down hard. Additionally, you can get a pen that perfectly fits your hand, which can be beneficial if you suffer from joint issues or arthritis.
3. Using a fountain pen is a special experience that is uniquely yours. Your pen, ink, and paper combination can be tailored to your exact needs.

4. Some great ways to use your pen are journaling, note-taking, letter writing, and handwriting practice.
5. Testing and playing with inks are great ways to get accustomed to the special properties in the ink you have selected. Get a spare sheet of paper, your ink, and a cotton swab, and have fun.

Frequently Asked Questions

Many of the questions below have been answered throughout the book. But if you would like a quick guide of the best tips, here are some very commonly asked questions about fountain pens.

1. What pen should I start with?

This is very subjective. You have to know what you like out of a pen and how you'll use it to make that decision perfectly. The best pens to start with are the Platinum Preppy, Pilot Metropolitan, Lamy Safari, and TWSBI ECO. These are some of the best starter pens that give you a great bang for your buck. The most expensive of the bunch is $35 USD. All of these are reliable writers and will last, even the four-dollar preppy.

Suppose you can find a physical shop to try out pens and see how they feel in your hand, the better. But these pens mentioned above are a great place to get started, so you can learn what you like out of a pen without shelling out a lot of money. Fountain pens can be an expensive hobby, but they don't have to be. Just make sure you get some ink with your pens, so you are good to write the moment you open the box.

2. Should I start with cartridges or bottled ink?

This is up to you and depends on the pen you have. If you have a cartridge converter pen, you can start with either type as long as you have a converter. If you have a piston filling pen, then you can only use bottled ink. Cartridges are more convenient, but bottled ink gives you more variety. The best way to decide where you want to start is to pick an ink regardless of what it comes in. Just pick an ink that you think you will like. If it only comes in cartridges, and the cartridge fits your pen, you start with it. If the ink only comes in bottled form, see if you can get a sample and try it out before committing to a full bottle.

3. Which way should I hold the pen?

Hold the pen by the grip section between your fingers like you would any other writing instrument. Get a feel for how comfortable it is in your hand and shift it forward or back if you need to. Some grip sections will be sculpted so that you hold the nib at the proper orientation. But if it doesn't, hold it with the shiny side to the sky. The metal nib should be facing up, not the feed.

Some pens can write with the nib facing down; this is called reverse writing. But many pens aren't designed to write this way, and the tip won't be able to make proper contact with the page. It's not recommended to write this way unless you have to.

If you're worried about losing your pen cap, see if you can snap it onto the back of your pen while you write. This is called 'posting your pen.' Not every pen is designed to post. So, if it doesn't snap or push into place easily, don't force it. You could crack the cap or body of the pen. This can make the pen more back weighted, so be careful that it doesn't make your pen uncomfortable. Most caps have clips that keep them from rolling away from you. There isn't much danger in putting it down on the table unless you're writing somewhere unfamiliar.

4. How important is the paper?

Very important. There's a whole section of this book on the paper alone. The paper you use will have a huge impact on how the pen writes. The same pen could write vastly different on average copy paper when compared to your favorite notebook. You have to keep in mind the qualities in the paper you're using when picking some pens. Many starter pens will have nibs that work well on any paper. Generally, the finer the nib you're writing with, the better chance you have of experiencing no issues with your writing. But if you want to use broader nibs or wetter writers, you have to look into the paper you're using to make sure that you are setting yourself up for success.

5. What does ink with shimmer mean?

Ink with shimmer means that there are tiny glitter particles suspended in the ink. When you write, these little particles will shine on the page. This shimmer adds a whole new depth to the color of the ink. When writing with shimmer ink, you want to get as much ink as possible down on the page to show off as much of the gorgeous color and glitter as you can. This ink is best used with a broad or wet writing nib.

6. What does ink with sheen mean?

Sheening ink is similar to shimmer, but instead of physical particulate, the ink is formulated specifically to shift to a different color depending on how much ink is on the page and the light's angle. This gives off a similar color variance to shimmer inks without dealing with cleaning out small pieces of glitter that have gotten stuck in your pen.

7. What does bulletproof ink mean?

Bulletproof is a term used by the Noodler's company to refer to its inks that are document proof. You'll find inks with the same properties being called document ink, archival ink, or document proof from other brands. Essentially, these inks are very durable and will not wash away so easily. They are waterproof and won't fade over time. Bulletproof or document proof inks are great for things like signing legal documents or checks. Water-based fountain pen ink can wash away over time. Document inks are meant to last.

8. Why are fountain pens so expensive?

When compared to your average box of ballpoint pens, a single fountain pen's cost can seem astronomical. What it boils down to most is that these pens are made to last. You're meant to use them again and again and not throw them out after each use. One fountain pen could outlast hundreds of disposable ballpoints if you use them long enough.

For others, you're paying for the craftsmanship that went into the pen. Some truly are tiny works of art that you can hold in your hand. Several things like the materials that went into the body, the intricate details carved into the finish, or the mixing of colors in the resin in the cap can all factor into the pen's cost. Aesthetics may not matter as much to you. It may be more worth it to you to invest the money in a pen that's made with better materials than one that's made with better-looking materials.

9. Should I invest in a more expensive pen upfront?

Not necessarily. Pens that cost more are not inherently better than cheaper pens. In some cases, you are paying for the craftsmanship and the pen's aesthetic, not the writing experience. At a certain point, the return-on-investment levels off when it comes to the overall writing experience. You're not always paying for quality or features; you're paying to write with a small work of art.

Many users feel like a pen should only have certain features such as gold nibs or built-in piston mechanisms after a certain price point. They won't pay premium prices for cartridge converter pens or steel nibs. This is a personal decision. Many beautifully crafted pens are cartridge converters, and some have steel nibs. It's up to you if the aesthetic is worth the price.

A good steel nib will always give you a better writing experience than a bad gold nib. Just because it costs more, it doesn't mean it's inherently better. It's just made of more expensive materials. Some starter pens cost more than others and will give you a different writing experience because of that. However, you don't need to spend large amounts of money upfront to get a better writing experience, especially if you don't yet know what you like. Experiment with cheaper pens in the sub $30-dollar category or invest in pens whose nibs you can swap without having to buy a whole new pen to decide what you like. Then determine if you want to spend more money on a gold nib pen or one with a more intricate aesthetic.

10. Why would I want more than one fountain pen?

That's entirely up to you. You may never want more than one fountain pen. You could strike gold with your first pen and never need to write with anything else. Many people end up with multiple fountain pens because it helps them figure out what they like out of a pen. You can't know what you like until you try that which you don't. Having multiple pens lets you try out different nib sizes, grips, pens of different weights and lengths, and everything in between. Trying out as much as you can for the least amount of money is a great way to figure out what you like out of a fountain pen.

For others, they want to have more than one fountain pen for the collection aspect. They like having all of something in every color it comes in. Collecting can be a fun part of the hobby that keeps it feeling fresh and new every time you open up the package with the newest pen to add to your collection. Collecting keeps you exploring, and it also gives you options as you could collect every color of a certain fountain pen but have all of them with different nib sizes. This way, you get a writing experience and pen you're familiar with, but it lets you play around with different nib sizes.

Variety is a big thing for a lot of fountain pen users. They not only have multiple fountain pens on hand at home, but they have multiple pens inked up at once. They like having the variety and choice to pick from different pens as the day goes on. They may not want to write with the same pen with the same ink all day for different purposes. They like having one for work notes, one for journaling, and one to write down their to-do list and grocery list with.

11. My pen won't write. What do I do?

There could be multiple reasons why your pen is not writing. The most common reason is that ink hasn't made its way through the feed. If you are using an ink cartridge for the first time, it will take time for the ink to work its way up through the feed to the nib. The best thing for this is to store the pen with the cap on and nib down. Gravity will help a little bit with getting the ink to work through the feed.

If you're using a converter that you filled separately before attaching it to the grip section, you can force some ink up through the feed. Slowly twist the piston knob on the converter until you see ink come up through the filler hole. Do this slowly so that you don't get ink all over yourself. Once the ink has worked its way to the feed, you're good to go.

Another reason that your pen isn't writing is that ink has dried in the feed, and it can't let fresh ink through. You have a few options. You can force ink through the feed like listed above, or you can dip your pen either in a glass of clean water or a bottle of ink. This extra moisture will loosen any dried bits that have gotten caught in the feed and help with the flow. The first few lines after doing this will be wetter than normal because the feed is so saturated. You may want to do some test lines before you get back to whatever writing you were doing.

If your pen isn't writing straight out of the box and you've forced ink through, your pen might need to be cleaned. It's recommended by most manufacturers that you give your pen a quick flush with clean water before inking it up for the first time. It's tempting to go right to writing with it, but there can be oils or dust from the manufacturing process stuck in the feed of your pen, impeding your flow. Please give it a quick flush; it doesn't have to be extensive, and then ink up your pen. Most of the time, this will fix any issues right out of the box.

Conclusion

Hopefully, this book has helped get you ready and excited to try out your first fountain pen. Choosing your first pen can be daunting as there are so many variables and options to consider. But as long as you remember the guidelines given to you, you can't go wrong. Pick something that feels right to you and that you will be excited to use every day. The right pen is the one that you pick up with enthusiasm and want to write with.

Remember to consider your handwriting style when choosing your pen and nib size. Pick a pen that will be comfortable in your hand, not too heavy and not too light. Make sure the pen you choose looks appealing to you. Your personal taste and aesthetic make all the difference when choosing the perfect pen.

Don't forget to go easy on the nib and write with gentle pressure. This one tip alone will make your writing experience that much better. You don't need to press down hard. Just guide the pen and let it do all the work.

Writing with a fountain pen should be an enjoyable experience. Pick up some ink with your pen so you can start experimenting with what you like. You'll soon find you'll have any and every excuse to write things down on paper instead of typing them into your phone or tablet. It's just plain fun. Now that you're armed with the tools and information you need go out and get started. Happy writing.

About the Expert

Lauren Traye is a life-long stationery enthusiast turned fountain pen fanatic. She's been collecting and using fountain pens for well over 2 years and enjoys spreading her knowledge to her friends, family, and anyone who will listen. Over the years, she's amassed a collection of over 20 pens, 18 bottles of ink, and countless ink samples. She's not running out of ink any time soon. Her preferred pen will have a fine or medium nib and will always be inked with something purple. When she's not tending to her fountain pen collection or roaming the internet for her next acquisition, she enjoys bullet journaling, reading, and spending time with her two adorable kittens.

HowExpert publishes quick 'how to' guides on all topics from A to Z by everyday experts. Visit HowExpert.com to learn more.

Recommended Resources

- HowExpert.com – Quick 'How To' Guides on All Topics from A to Z by Everyday Experts.
- HowExpert.com/free – Free HowExpert Email Newsletter.
- HowExpert.com/books – HowExpert Books
- HowExpert.com/courses – HowExpert Courses
- HowExpert.com/clothing – HowExpert Clothing
- HowExpert.com/membership – HowExpert Membership Site
- HowExpert.com/affiliates – HowExpert Affiliate Program
- HowExpert.com/jobs – HowExpert Jobs
- HowExpert.com/writers – Write About Your #1 Passion/Knowledge/Expertise & Become a HowExpert Author.
- HowExpert.com/resources – Additional HowExpert Recommended Resources
- YouTube.com/HowExpert – Subscribe to HowExpert YouTube.
- Instagram.com/HowExpert – Follow HowExpert on Instagram.
- Facebook.com/HowExpert – Follow HowExpert on Facebook.

Printed in Great Britain
by Amazon